OSPREY AVIATION ELITE • 4

Lentolaivue 24

MT-441

SERIES EDITOR: TONY HOLMES

OSPREY AVIATION ELITE • 4

Lentolaivue 24

Kari Stenman & Kalevi Keskinen

OSPREY
AVIATION

Front cover
On 24 January 1943 at 1440 hrs, Capt Jorma Sarvanto took off from Suulajärvi with five other Brewster Model 239s from 1/LeLv 24 and headed east for a 'free hunt' over the Gulf of Finland. The Soviets had a huge naval and air base on the island of Kronstadt, just east of Leningrad, and the Finnish pilots knew their sortie would be opposed. As the aircraft approached Kronstadt, 1Lt Hans Wind (leading the second pair of Brewsters) observed six Il-2s, escorted by two I-16 'Ratas' and seven 'Spitfires' (actually LaGG-3s) below the Finnish fighters. Moments later he spotted five Pe-2s and a further 13 I-16s. All these aircraft were heading in an easterly direction, slowly descending towards the Soviet airfields at Kronstadt and Oranienbaum. With the sun behind them, Wind (in BW-393) and his wingman attacked the two 'Ratas' closest to them, the former jumping his foe at an altitude of less than 100 metres above the ground. The I-16 was mortally damaged by a short burst of fire from Wind's Brewster, and it crashed into a forest just west of the runway at Kronstadt. Its demise (Wind's 15th kill of the war) was witnessed both by his wingman, WO Viktor Pyötsiä, and the Finnish ground post at Seivästö. Both pilots then climbed back up to altitude and joined the battle raging overhead with the I-16-escorted Pe-2 formation. Wind duly fired at two other 'Ratas' and two Pe-2s before his guns jammed and he was forced to head for home. The Finnish pilots reported that their Russian opponents were flying aircraft painted in green summer camouflage, despite it being the middle of winter! (*Cover Painting by Jim Laurier*)

Back cover
On 1 June 1943 *Lentolaivue* 24's 3rd Flight leader, Capt Joppe Karhunen, was appointed commander of the whole unit. Marking this occasion, he poses on the tailplane of his Brewster Model 239 (BW-366) at Suulajärvi. The fighter's fin shows the Mannerheim Cross-winner's full tally of 31 aerial victories, the last of which was scored on 4 May 1943 (*SA-kuva*)

First published in Great Britain in 2001 by Osprey Publishing
Elms Court, Chapel Way, Botley, Oxford, OX2 9LP
E-mail: info@ospreypublishing.com

ISBN 1 84176 262 8

Edited by Tony Holmes
Page design by Mark Holt
Cover Artwork by Jim Laurier
Aircraft Profiles by John Weal
Origination by Grasmere Digital Imaging, Leeds, UK
Printed through Bookbuilders, Hong Kong

01 02 03 04 10 9 8 7 6 5 4 3 2 1

ACKNOWLEDGEMENTS
The Authors wish to thank Carl-Fredrik Geust for providing details of Soviet units and operations on the Finnish front, this information having been sourced from original documentation found in Russian archives kept near Moscow and St Petersburg

Previous pages
A pair of Bf 109G-6s from 3/HLeLv 24 are seen between sorties at Lappeenranta on 3 July 1944. MT-441/'Yellow 1' (left) was flown by 1Lt Ahti Laitinen until he was captured on 29 June 1944 after bailing out of battle-damaged Bf 109G-6 MT-439. Laitinen had, up to that point, flown just 75 sorties and claimed ten kills – six of them in MT-441. The second *Gustav* is MT-476/'Yellow 7', which was assigned to MSgt Nils Katajainen for a mere 48 hours! On 5 July 1944 he used it to down a Yak-9, but was in turn wounded and forced to crash-land at 500 km/h – miraculously, he survived this incident. Nils Katajainen scored 35.5 kills during the course of 196 sorties, and received the Mannerheim Cross (Finland's highest military medal) on 21 December 1944 (*SA-kuva*)

CONTENTS

HUMBLE BEGINNINGS

Utti air base in south-eastern Finland has long been recognised as the 'cradle of the fighter pilot' following its establishment in June 1918 – six months after the Finnish declaration of independence from then Imperial Russia. As the only airfield for land-based aircraft in the country, Utti was home to all flying training and frontline fighter units for over a decade. Indeed, it was not until 1929 that a second military airfield was built close to Viipuri at Suur-Merijoki. By this time the Finnish Air Force had established an effective maritime flying arm thanks to the instruction of a British military team headed by Maj Gen Walter Kirke.

Flying boats and floatplanes continued to dominate Finnish military aviation until the mid-1930s, and this was one of the primary reasons why the air force boasted such a small fighter arm at the start of World War 2. Another principal cause was that much of the meagre pre-war Finnish military budget had been invested in two 3900-ton light cruisers purchased to boost the image of the navy! These warships subsequently saw very little use during World War 2, for they were always vulnerable to attack from the air when at sea due to the dense archipelago of the Gulf of Finland.

The modernisation of the air force finally commenced on 15 July 1933 when a series of new air stations were created, and each one allocated one or two previously-established squadrons. Utti was designated *Lentoasema* 1, and its squadrons were also numbered – the army co-operation unit became *Lentolaivue* (LLv) 10 and the fighter outfit *Lentolaivue* 24.

LLv 24 was then equipped with licence-built Gloster Gamecock IIIs, which it had received three years earlier. These had replaced French Gourdou-Leseurre GL-21s (20 of which were purchased in 1923) and British Martinsyde F.4 Buzzards (15 acquired).

FIGHTER TACTICS

The Gamecocks gave great service, and in 1934 LLv 24's commanding officer, Maj Richard Lorentz, used them to experiment with new fighter tactics which saw the traditional lead aircraft and two wingmen formation reduced to just a 'fighting pair'. This proved to be both more flexible and better suited to most tactical conditions, and could be easily increased to a four-aircraft 'swarm' (dubbed a 'finger-four') when the need arose.

Further tactical improvements within LLv 24 were implemented following Capt Gustaf Magnusson's arrival as CO on 21 November 1938. Prior to his appointment, he had visited other air arms in Europe, including a three-month spell with the Luftwaffe's JG 132 *'Richthofen'*. A number of its pilots had recently seen action in the Spanish Civil War, and Magnusson received valuable information on how best to destroy Soviet Tupolev SB bombers and Polikarpov I-15bis and I-16 fighters. The Germans had also abandoned the three-aeroplane formation in favour of the 'finger-four', and this convinced senior officers in the Finnish Air Force of the soundness of LLv 24's fighting formations and basic tactics.

Yet poor funding continued to restrict pre-war fighter pilots from receiving anything but elementary flying training prior to reaching the

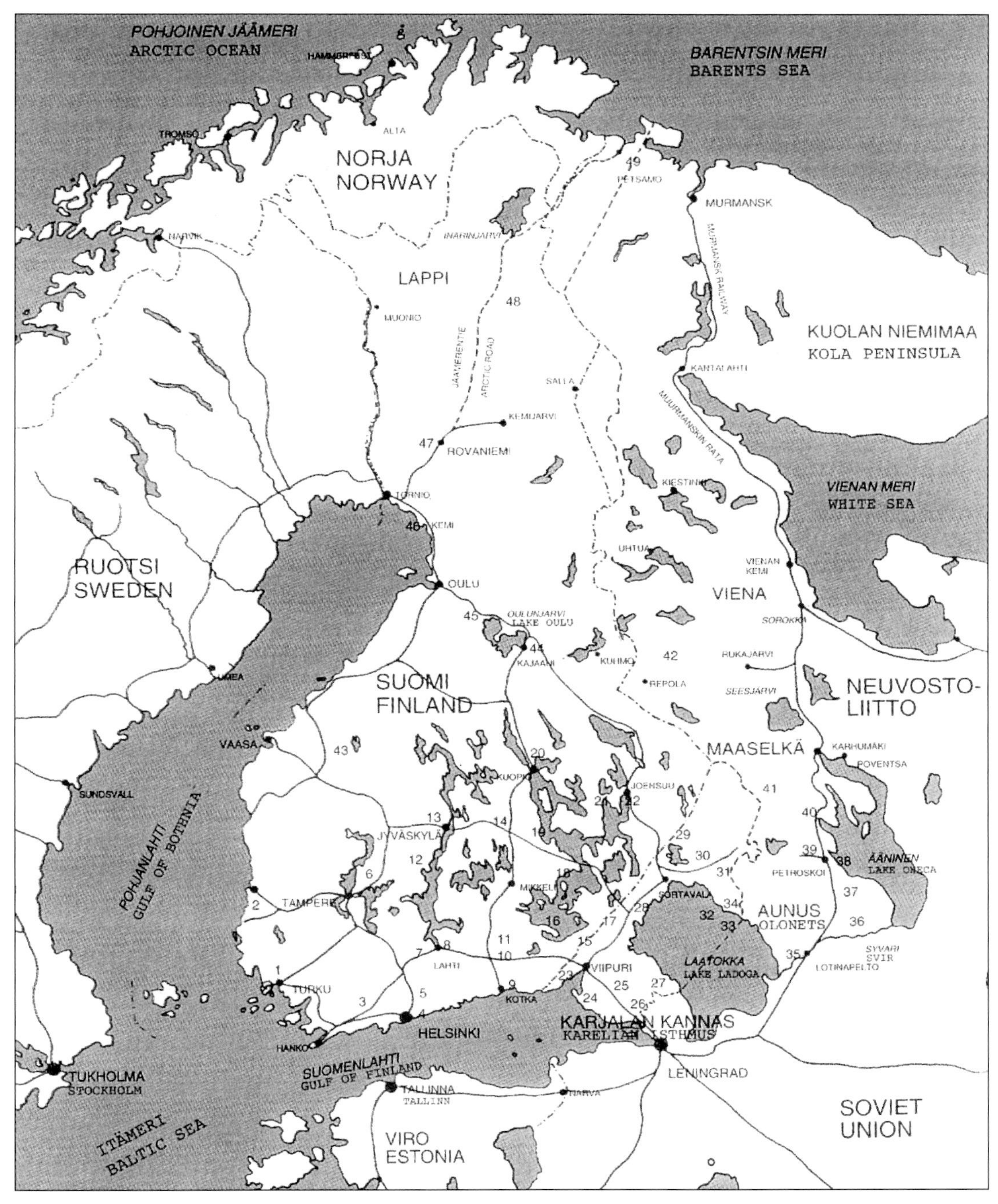

AIRFIELDS

1. Turku
2. Pori
3. Nummela
4. Malmi
5. Hyvinkää
6. Siikakangas
7. Hollola
8. Vesivehmaa
9. Kymi
10. Utti
11. Selänpää
12. Kuorevesi
13. Luonetjärvi
14. Naarajärvi
15. Lappeenranta
16. Taipalsaari
17. Immola
18. Rantasalmi
19. Joroinen
20. Rissala
21. Onttola
22. Joensuu
23. Suur-Merijoki
24. Römpötti
25. Heinjoki
26. Suulajärvi
27. Kilpasilta
28. Mensuvaara
29. Värtsilä
30. Suistamo
31. Uomaa
32. Mantsi
33. Lunkula
34. Karkunranta
35. Nurmoila
36. Latva
37. Derevjannoje
38. Solomanni
39. Viitana
40. Kontupohja
41. Hirvas
42. Tiiksjärvi
43. Kauhava
44. Paltamo
45. Vaala
46. Kemi
47. Rovaniemi
48. Vuotso
49. Petsamo

frontline. And even once they had arrived at LLv 24, pilots found that advanced training rarely explored more than three types of attack, for it had been discovered that such a number was sufficient to down a bomber – the Finnish fighter pilot's principal target. The trio of attacks (and associated aerial gunnery) adopted by LLv 24 were thoroughly rehearsed, and this style of training suited both the air force and the Finnish treasury.

When practising aerial attacks, guns were set to converge at 150 metres, but pilots were trained to hold their fire until just 50 metres away from the target. Such closeness brought with it certain risks, but these were deemed to be outweighed by two major advantages: 1) you were too close for any defensive fire to be aimed with any accuracy, and 2) you could not miss!

When war broke out Magnusson ordered that fighter duelling was to be avoided, for LLv 24's Fokker D.XXIs could not turn with Russian I-15s and I-16s. However, the Dutch design was a capable interceptor, possessing a good rate of climb and the ability to dive away from the enemy.

On 1 January 1938 the air stations were replaced by flying regiments, and newly-delivered D.XXIs were issued to both of *Lentorykmentti* 2's fighter units, LLv 24 and 26. An intensive training period ensued, and both FR-79 and FR-88 (and their pilots) were lost in accidents.

FOKKER D.XXI

The Finnish Air Force initially became involved with Fokker designs when it purchased a number of C.V army co-operation aircraft in the early 1930s. These were in turn replaced in the spring of 1936 by the C.X, these being bought as part of the new five-year plan which also called for the acquisition of 27 'interceptors' to equip three squadrons.

Fokker had offered the Finns its new low-wing monoplane D.XXI fighter for export, and they duly became the aircraft's first customer on 18 November 1936. Seven aircraft were ordered, and a licence was also acquired to produce double this amount at the *Valtion Lentokonetehdas* (State Aircraft Factory). Bearing Finnish serial numbers FR-76 to FR-82 (series I) inclusive, the Dutch aircraft arrived in Amsterdam on 12 October 1937 and were shipped on to Finland in crates. Each D.XXI cost 1.1 million Finn marks apiece without an engine.

On 7 June 1937 an order was placed with the *Valtion Lentokonetehdas* for 14 series II aircraft (serialled FR-83 to FR-96). Completed between 11 November 1938 and 18 March 1939, these D.XXIs cost just half the price of the Dutch examples.

An open licence had also been obtained on 15 June 1937, which allowed 21 series III aircraft to be built between 16 March and 27 July 1939 – just in time for the Winter War. These fighters were serialled FR-97 to FR-117 inclusive.

The first D.XXI issued to LLv 24 was FR-76, which is seen here in front of LA 1 hangar No 2 on 16 December 1937. Its unique 20 mm Oerlikon cannon are clearly visible under the fighter's wings. On 29 January 1940 2Lt Olli Puhakka downed a DB-3M from 43.DBAP (long-range bomber aviation regiment) with just 18 rounds fired from a distance of over 500 metres (see chapter two). Within a week machine guns had replaced the cannon, allowing the fighter to comply with the rest of the unit's machines (*Finnish Air Force*)

WINTER WAR

Germany launched its long-awaited attack on Poland in the early hours of 1 September 1939, and seized the western areas of the country within three weeks. According to the Ribbentrop-Molotov non-aggression pact signed between the Nazi regime and the Soviet government just weeks prior to the invasion, Poland was carved up between the two countries. A secret clause appended to the pact left the Baltic countries, and Finland, to the Soviet Union, which now also occupied the eastern provinces of Poland. At the same time the communists demanded air and naval bases from then independent Estonia, Latvia and Lithuania, which had to comply due to the weak state of their armed forces. The Soviets then turned their attention to Finland.

Having enjoyed success with the Baltic states through the application of massive diplomatic pressure, the communists initially tried to acquire military bases from Finland through identical means. Realising that the Germans would eventually attempt to invade at some point in the near future, the Soviets wished to shift their border further west from Leningrad. In return for giving up this land, Finland would get twice the area of wilderness further north in Soviet Karelia. All communist overtures were disguised under the 'safety of Leningrad' banner, which the Finns correctly suspected was a cover for the total conquest of their country. The Finnish government flatly refused this insult to its sovereignty.

Following the receipt of this answer, the Soviet Union annulled the non-aggression pact signed with Finland in 1932 and launched an invasion on 30 November 1939. So started the 'David and Goliath' struggle that was the Winter War.

RE-EQUIPMENT

In 1937 the Finnish Air Force had issued a five-year development plan which called primarily for the acquisition of 'interceptors'. It had been

Neatly lined up at an airshow held at *Lentoasema* (Air Station) 5 – previously known as Suur-Merijoki – on 3 August 1935, these Gloster Gamecock IIIs of LLv 24 provided the bulk of Finland's frontline fighter defence for much of the 1930s. The unit was based at Utti – home of *Lentoasema* 1 – at the time. On 1 January 1938 aviation regiments were formed to replace the air stations, and Utti became the home of *Lentorykmentti* 2. Some 17 Gamecocks saw exactly a decade of service with the Finnish Air Force from 1929 through to early 1939, when they were replaced by Fokker D.XXIs (*Finnish Air Force*)

LLv 24's Dutch-built D.XXI FR-80 is seen with its tail hoisted up on a trellis at Utti on 30 August 1938. Once in this configuration, the fighter became the subject for a series of recognition photographs taken by an air force photographer. All seven machines imported directly from Holland featured dark brown upper surfaces and aluminium dope undersides, while the Finnish-built D.XXIs were painted standard air force olive green and light grey. FR-80 was subsequently shot down over Helsinki by an I-16 from 25.IAP on 19 February 1940. Its pilot, 1Lt Erhard Frijs, was killed (*Finnish Air Force*)

correctly deduced that any enemy attacking Finland would rely heavily on the large-scale use of bombers, without fighter escort. With only limited funds available, the Finns would have to procure these interceptors from sources other than the major European powers, which could not spare military aircraft in the growing climate of political tension. And because of the difficulties encountered in obtaining modern fighters, the air force had only received two-thirds of its aircraft when the Red Army attacked.

The Soviets had mustered 450,000 men in close to 20 divisions along the Finnish border. These troops were supported by 2000 artillery pieces, 2000 tanks and 3253 aircraft – the latter would fly an average of 1000 sorties per day throughout the war. In opposition, the Finns committed five divisions, 300 guns, 20 tanks and 114 serviceable aircraft to the main front at the Karelian Isthmus .

All Finnish fighter defences were controlled by *Lentorykmentti* 2, which was commanded by ex-LLv 24 CO, Lt Col Richard Lorentz. When it was realised that the political situation was deteriorating towards war, his old unit was one of two fighter squadrons that Lorentz was ordered to disperse to new airfields away from their usual bases. The D.XXI-equipped LLv 24 received a further ten Fokker fighters – and pilots – from LLv 26 (the other squadron to move) on 26 November 1939, leaving the latter unit with just ten obsolete Bristol Bulldogs. Following this switch, Finland's entire frontline fighter force consisted of just LLv 24! Its commanding officer, Capt 'Eka' Magnusson, would later show exceptional tactical and personal leadership as he led his 35 Fokkers into battle over and over again.

LLv 24 was ready to be dispersed along the frontline of south-eastern Finland, its well-trained and highly motivated pilots being split into the following five flights;

***Lentolaivue* 24 on 30 November 1939**

Commander	Capt Gustaf Magnusson, with HQ at Immola
1st Flight	Capt Eino Carlsson at Immola with six D.XXIs
2nd Flight	1Lt Jaakko Vuorela at Suur-Merijoki with six D.XXIs
3rd Flight	1Lt Eino Luukkanen at Immola with six D.XXIs
4th Flight	Capt Gustaf Magnusson at Immola with seven D.XXIs
5th Flight	1Lt Leo Ahola at Immola with ten D.XXIs

LLv 24's task was to protect the traffic junctions in south-eastern Finland, and prevent attacks either on or through the Karelian Isthmus. Although the D.XXI lacked speed and heavy armament, it was an ideal interceptor, with a good rate of climb and excellent diving characteristics. And with fixed ski-landing gear, it could be flown from very austere bases.

FIRST ENCOUNTERS

On 30 November 1939 the Soviets sortied 200 bombers against both towns and bases in southern Finland, and due to poor weather the interceptors failed to engage the Soviet force. Helsinki was among the locations to be bombed when eight Ilyushin DB-3s of 1.MTAP KBF (a mine-torpedo aviation regiment of the Red Banner Baltic Fleet) caused close to 300 civilian casualties, including nearly 100 dead. This horrendous loss of life brought the Finns much sympathy from countries across the globe.

Ironically, recent research in the Russian archives has revealed that the bombing of Helsinki's city centre took place by mistake, with the actual target being the port and oil storage facility at Herttoniemi, several kilometres further east.

On the morning of 1 December the Red Air Force sent out a wave 250 unescorted bombers to hit the same targets that had been successfully attacked 24 hours earlier. The Finnish fighter pilots were determined to effect an interception following the disappointment of the previous day, and LLv 24 scrambled its Fokkers in pairs, led by the CO Capt Magnusson in his assigned D.XXI, FR-99. A total of 59 sorties were flown by the unit, whose pilots claimed 11 Tupolev SB bombers destroyed in the Viipuri-Lappeenranta area – eight from 41.SBAP (fast bomber aviation regiment) and three from 24.SBAP.

LLv 24's inspirational commander, Gustaf Erik 'Eka' Magnusson, was promoted to major on 6 December 1939. To mark the occasion, he gave a speech to his men at Immola – where this photograph was taken. Behind him are D.XXIs FR-105 and FR-106 (*I Juutilainen*)

The first fell at 1205 hrs to 2nd Flight leader 1Lt Jaakko Vuorela (in FR-86), and the last at 1440 hrs to 5th Flight boss 1Lt Leo Ahola (in FR-113). Vuorela also claimed a second SB to become the first Finnish pilot to claim more than one kill. The remaining victories were scored by Capt Gustaf Magnusson, 1Lts Eino Luukkanen and Jussi Räty, 2Lt Pekka Kokko and Sgts Lasse Heikinaro, Lauri Nissinen, Lauri Rautakorpi and Kelpo Virta.

Sadly, no combat reports exist from these first encounters, for such forms were not made available for another three weeks! However, Capt 'Eka' Magnusson insisted that every pilot involved in combat on this day should make a record of his experiences. His own note reads;

'1/12/39 at 1410-1445 hours. We took off following an announcement that a Soviet bomber formation was approaching Imatra. We met the formation above Imatra. I attacked the bomber flying on the extreme right wing of the formation, shooting first along its fuselage. When my firing did not seem to have any effect, I aimed at the starboard engine instead, which started to smoke after a few bursts.

'I then had to interrupt my attack since the bomber to the left of my target had reduced his speed so that he was about 70 metres off my port side, with his dorsal gunner firing at me. I too slowed down and attacked this aeroplane instead, shooting it down in flames – the bomber continued to burn after hitting the ground. Since the squadron had only normal bullets and tracer rounds, it was impossible to gain results with a small amount of ammunition – hence I fired 1200 rounds. I was flying FR-99.'

Capt Magnusson's clinical attack proved to his men that the tactics taught to the pilots of LLv 24 throughout the 1930s would allow them to engage a numerically superior enemy with confidence.

The unit suffered its first loss of the conflict shortly after its opening successes when 'friendly' anti-aircraft fire at Viipuri downed FR-77, killing Sgt Matti Kukkonen. A further loss occurred when the squadron 'hack', de Havilland Moth MO-111, was struck by a bomb at Immola.

Possessing such few fighters, the Finns had wisely mobilised their forces several weeks prior to the Soviet invasion in an effort to avoid the D.XXIs being caught on the ground at the principal air force bases. Following the opening 48 hours of the Winter War, Maj (from 6 December) Magnusson also made operational changes in an effort to shorten LLv 24's chain of command. These saw the 2nd and 5th Flights combined as Detachment Vuorela and transferred, on 9 December, to Lappeenranta, whilst the 1st Flight moved to Mensuvaara. On 18 December the squadron was relieved of defensive duties in western Finland, allowing it to concentrate fully on supporting the army on the Karelian Isthmus.

Poor weather and snow fall stopped all operational flying until 19 December, when LLv 24 flew 58 sorties over the Karelian Isthmus and engaged the enemy in combat on 22 occasions between 1050 and 1520 hrs. The Soviets lost seven SBs (six from 44.SBAP) and five Ilyushin DB-3s from other regiments, SSgt Kelpo Virta (in FR-84) being the first pilot to engage Soviet aircraft on this day. He had encountered I-16 fighters, from 25.IAP, rather than bombers, however;

'I was flying as a wingman in a three-aeroplane division, patrolling south-east of Viipuri (Vyborg), at 1010 hrs when we observed nine I-16s ahead and 500 metres below us. The enemy fighters formed a circle and started to gain altitude. We did the same. Sgt Nissinen then dived down

Pilots from LLv 24 pose in front of FR-110 in January 1940. They are, from left to right, Sgt Martti Alho, 2Lt Tapani Harmaja, 1Lts Jussi Räty and Veikko Karu, Maj Gustaf Magnusson, MSgt Sakari Ikonen, WO Viktor Pyötsiä (stood behind Ikonen), 2Lt Iikka Törrönen, 1Lt Per-Erik Sovelius (stood behind Törrönen) and an unnamed war correspondent. Tapani Harmaja was killed in action on 1 February 1940 and Sakari Ikonen wounded nine days later (*J Sarvanto*)

Its hot engines still steaming in the snow, this Tupolev SB-2M-100 was downed by 1Lt Jussi Räty of 4/LLv 24 in FR-115 on 1 December 1939 at Imatra. 'Yellow 9' (construction number 20/101) belonged to 41.SBAP (fast bomber aviation regiment), and of its crew of three, pilot Lt Zhorka Tanklayev and navigator 1Lt Viktor Demchinskiy were captured, while gunner Cpl Sergei Korotkov was killed (*SA-kuva*)

SSgt Lauri Nissinen leans on his D.XXI whilst posing for the camera in early 1940. A member of the 5th Flight, he had claimed an I-16 damaged over the Karelian Isthmus on 23 December 1939 whilst flying FR-98. According to his combat report, the Polikarpov fighter bore the tactical number '228' – unusually high for the Red Air Force, which usually employed two-digit numbers (*E Luukkanen*)

to attack them, and I followed him down with the sun behind me. I attacked an enemy aeroplane which was a little apart from the others, firing a couple of bursts from 50 metres. The tracers seemed to hit but without any result, so I repeated my attack and shot at him from very close range. The enemy aircraft caught fire after two bursts and I pulled up. I looked around and saw none of my squadronmates.

'I then attacked a second aircraft, firing short bursts from behind it at distances ranging from between 50 to 20 metres. By this stage we were both diving at a 45-degree angle. The aircraft started to emit white smoke, and it increased the angle of its dive until it was almost vertical. I then spotted three other enemy aeroplanes behind me, so I banked steeply, went into a spiral dive, levelled off and fired at an I-16 coming towards me. I ended up below my enemies, so I dived for the ground and came home.

'The battle started over the northern end of Lake Muolaanjärvi at 1015 hrs and ended at the northern end of Lake Suulajärvi at 1025 hrs. I fired 650 rounds, and my starboard wing gun was out of order. I was flying FR-84.'

This photograph proves that Nissinen's damage claim needed to be upgraded to a full kill! The wreckage of 'Red 228' was photographed at Muolaa on 28 December 1939 following its recovery by the Finns. The downed I-16 proved to be a 25.IAP (fighter aviation regiment) machine, and its pilot, 1Lt Iosif Kovalkov, was made a PoW (*Finnish Air Force*)

Troops on the ground had seen two I-16s crash nearby, and Virta was duly credited with having destroyed both aircraft. LLv 24's adjutant interviewed every pilot engaged in combat on this day, and summarised events in combat report style. 1Lt 'Pelle' Sovelius, flying FR-92, recounted;

'On 19/12/39, from 0955 to 1105 hrs, I was leading the third pair on an air combat patrol, with Sgt Ikonen on my wing. After being scrambled and climbing out over Antrea, we were ordered by radio to change course and head towards the south-west. When nearing Kämärä, I observed a seven-aeroplane SB formation and commenced the chase. The SBs initially flew towards the south-west, although they then turned due south. Failing to gain on them, we then observed three more SBs flying in roughly the same direction.

'Sgt Ikonen quickly got in behind the aeroplane on the starboard wing of the formation and shot it down in flames from very close range at an altitude of 2000 metres over Kipinola. I tried to get in behind the bomber to port, but did not have enough speed. I observed another three SBs heading south-west and tried to get after them, but they quickly pulled away. These aeroplanes were dispensing leaflets.

'During the chase I observed three more SBs a little below me, heading southwards. I picked out the port wingman as my target, and initially shot into its rear fuselage to silence the dorsal gunner. After this I aimed my guns at the port engine, which began to smoke and finally caught fire. The aeroplane fell away to starboard and dived into the sea close to Seivästö, about ten kilometres from the coast.

'I then fired at the starboard wingman, whose right engine started to pour smoke, but it steadfastly remained in formation with the lead aeroplane. I performed the return flight at heights of between 3000 and 5000 metres, and when about five kilometres above the land I observed flak. Whilst flying along the south coast of Lake Muolaanjärvi, two I-16s managed to take me completely by surprise by diving out of the sun. I woke up when bullets smashed into my aeroplane.

'I instantly pulled in towards them, but quickly realised that the I-16 was more agile than the Fokker. I tried to tighten my turns, but only managed to get the enemy aircraft into my sights on one occasion, firing a short burst. It was at this point that I noticed that I only had ammunition left in one gun. After trying to turn as tightly as possible, I stalled the fighter and it fell into a spin. I continued to perform all manner of evasive actions until nearing the ground, where I managed to shake off my pursuers. I was then nearby Heinjoki. Both I-16s had attacked simultaneously, and in order to avoid a collision they had not got in straight behind me, instead being forced to shoot with a small deflection. Judging by their tracers, the I-16s seemed to continue firing even when I was clearly out of their sights.

'I landed with two holes in my aircraft. One round had struck the tailplane and the other had gone in through the machine gun compression bottle hatch and out the bottom of the fuselage.'

Four days later the luckless 44.SBAP was again badly mauled by Magnusson's Fokkers, losing six SBs between 1100 and 1200 hrs over the Karelian Isthmus. 1Lt Jorma Sarvanto claimed two in FR-97, while Maj 'Eka' Magnusson, 1Lt 'Joppe' Karhunen and WO 'Pappa' Turkka each downed one apiece – the destruction of the sixth bomber was shared. Here, Magnusson describes shooting down his bomber in FR-99;

'Sgt Kinnunen, flying on the left flank of the formation, observed nine SBs over Vuoksenranta. I dived after Sgt Kinnunen, who then pulled away thinking that I was an I-16.

'I continued after the formation and intercepted it over Kiviniemi. I chose the rearmost bomber on the left flank as my target, firing firstly at its starboard engine, which started to smoke. I then hit the port engine and it too burst into flames. The aeroplane then began to descend.

'Tactically, the enemy unit worked well, lowering their landing gear simultaneously for speed reduction. The aircraft next to the one under attack also reduced his speed so as to improve the rear gunner's field of fire.

'At 1200 hrs the aircraft hit the ground at Lempaalanjärvi.'

Pilots from 4/LLv 24 gather in front of Maj Magnusson's FR-99 'Black 1' at Joutseno in January 1940. They are, from left to right, Sgt Martti Alho, Danish volunteer 1Lt Fritz Rasmussen, 2Lt Tapani Harmaja and deputy flight leader 1Lt 'Pelle' Sovelius. Fritz Rasmussen was killed in action (in FR-81) by an I-16 from 25.IAP on 2 February (*J Sarvanto*)

Seen on 6 January 1940 sat in FR-97, 4/LLv 24's 1Lt Jorma 'Zamba' Sarvanto has good reason to be smiling, for he had just shot down six DB-3 bombers in four minutes south of Utti whilst flying this very fighter. Like most other Finnish aces, Sarvanto later saw action in Brewsters, scoring four victories to raise his final tally to 17 in 251 missions (*J Sarvanto*)

Some 21 engagements occurred during the course of the day, and aside from the SBs destroyed, two I-16s from both 7. and 64.IAP were shot down. Sgt Pentti Tilli, flying FR-103, accounted for the former pair and 1Lt Urho Nieminen and 2Lt Heikki Ilveskorpi the latter. A Soviet I-16 from 25.IAP did succeed in shooting down Sgt Tauno Kaarma, however, who was injured when FR-111 crash-landed at Lyykylänärvi.

The action continued on Christmas Day, when the Fokkers destroyed two SB bombers from 6.DBAP (Long-range Bomber Aviation Regiment) over the Karelian Isthmus.

The 25th also saw the 3rd Flight strengthened and redesignated Detachment Luukkanen, whereupon it was transferred to Värtsilä to support the troops on the northern coast of Lake Ladoga. Upon its arrival, the detachment immediately downed four SBs of 18.BAP, with 1Lt Jorma Karhunen in FR-112 and Sgt Toivo Vuorimaa in FR-93 claiming two apiece. On the 27th the Soviets lost three more SBs (from 2.SBAP) over the Karelian Isthmus, whilst WO 'Isä-Vikki' Pyötsiä downed two I-15bis fighters north of Lake Ladoga in FR-110.

Immola was frequently a target for Soviet bombers at this stage in the war, so the headquarters, 1st and 4th Flights moved to Joutseno on 28 December, and then on to Utti four days later – by this stage in the conflict Finnish military intelligence had discovered that bomber crews navigated their way to targets in Finland by following the railway network.

On New Year's Eve, Sgt Ilmari Juutilainen (a member of Detachment Luukkanen, and future ranking Finnish ace with 94 kills), flying FR-106, shot a lone I-16 off the tail of 1Lt Karhunen during an engagement over the northern shore of Lake Ladoga. This was the unit's final kill of 1939.

LLv 24's pilots had scored steadily during their first month of operations, claiming 54 aircraft downed for just one D.XXI lost in combat and another Fokker fighter damaged.

Four pilots from LLv 24's 3rd Flight smile for the camera at Ruokolahti in late February 1940. Sitting on the D.XXI's wheel spat is flight leader Capt Eino Luukkanen, and standing next to him is MSgt Ilmari Juutilainen. Sat on the wing are Sgts Jalo Dahl and Martti Alho (*E Luukkanen*)

By this time the similarly outnumbered Finnish ground forces had halted the Soviet advances along the 1000-mile frontline. The extremely cold winter, where temperatures often fell to -30° centigrade (and on several days plummeted to -40°C), favoured the Finnish defenders.

6 JANUARY 1940

During the morning of 6 January, 17 DB-3Ms of 6.DBAP took off in two waves from Estonia to bomb Kuopio, in eastern Finland. The first nine Ilyushins attacked their target as planned, but the second formation of eight drifted too far west and crossed the Gulf of Finland south of Utti. Based nearby was 4/LLv 24, which had 1Lt Sovelius (in FR-92) up on patrol. He attacked the DB-3Ms at 1010 hrs at a height of 3000 metres, downing the outer aircraft on the far left side of the formation.

The remaining seven bombers continued to Kuopio, where they released their bombs to little effect before returning home along the same route, which followed a railway line. 1Lt Sarvanto had, meanwhile, taken off in order meet the DB-3s on their return journey, and in his post-war memoirs, he described the famous four-minute battle which ensued;

'The clouds over Utti had disappeared and the sun gleamed from the light bellies of the marvellous looking row of bombers. I counted them to be seven. On the left flew an echelon of three and to the right four almost in a row. The distance between the aeroplanes was hardly one aircraft.

'I banked to the right and headed south, continuing to climb. For a moment I was in the sights of the nose gunners, but facing the sun, they obviously did not see me. When I reached the altitude of the bombers, I was already 500 metres behind them. At full power, I started the chase and selected the one at the extreme left of the formation, although the bomber third from the left was further behind the others, and the fire from its rear gunner felt dangerous. At a distance of 300 metres it banged unpleasantly into my aeroplane – I had flown into a stream of bullets.

'I opened fire at 20 metres with a short burst to the fuselage of the machine on the left. The tracers seemed to hit the target, and I quickly silenced the bomber's rear gunner. I took aim again at the right engines of both bombers in the formation, and with light touches on the trigger, both went down in flames. I cheered, and then aligned my Fokker up with the bombers on the opposite side of the formation. Attacking as I had before, I set the engines of one bomber alight, before turning to the next aircraft in the formation, hitting it with gun fire at very close range. This aeroplane also burst into flames soon after I had hit it with two or three very short bursts. I then saw the first aircraft that I had attacked on the right side of the formation diving as a fireball towards the ground.

'I now set myself the goal of destroying all the remaining bombers in the formation. Some fell away like burning pages of a book after I had fired at them, whilst others pulled up steeply following the incapacitation of their pilot. The reddish January sun shone through the haze towards me throughout the engagement, except when the dark smoke of the burning aeroplanes cast a shadow across it.

'The penultimate bomber was much tougher than the others to shoot down, for my wing guns were probably empty by then. It did, however, finally catch fire, and I in turn went after the last one. Its rear gunner had been silent for quite some time, and I went in very close. I aimed at the engine and pulled the trigger. The guns were quiet! I made a couple of charging attempts but without any result. I had run out of ammunition, and the only thing to do was to return home.'

The sole surviving DB-3 was chased down by 1Lt Sovelius, who had landed and had his fighter refuelled and rearmed since intercepting the bombers on the way to the target. He caught the fleeing Ilyushin as it flew out over the Gulf of Finland, sending it down in flames.

Sarvanto returned to base to be credited with the destruction of the remaining six DB-3s in the formation, for all of them had crashed between Utti and Tavastila – a distance of 30 kms. He thus became Finland's premier ace in an action that lasted just four minutes. His D.XXI (FR-97)

Air- and groundcrews from Detachment Luukkanen gather at Värtsilä on Christmas Eve 1939. They are, from left to right, an unnamed assistant mechanic, an unnamed armourer, mechanic P Hannula, mechanic J Paajanen, Sgt I Juutilainen, assistant mechanic T Karhu, SSgt P Tilli, mechanic P Heino, 1Lt T Huhanantti, mechanic V Eve, flight leader 1Lt E Luukkanen, mechanic U Raunio, 1Lt J Karhunen, assistant mechanic K Pyötsiä and mechanic E Horppu. Both Pentti Tilli and Tatu Huhanantti fell victim to I-16s during the Winter War (*E Luukkanen*)

had received 23 hits during the engagement, but none were serious, and it was flown to the repair facility. Once news of the action was released, foreign pressmen showed much interest in Jorma Sarvanto, for nothing like this had occurred in Europe up to this point in the war.

On 7 January heavy snow began to fall throughout the region, preventing virtually all flying on both sides of the front for over a week. Taking advantage of the conditions, the 1st and 4th Flights returned to Joutseno.

Ten days later, on the 17th, ten D.XXIs that had scrambled at 1355 hrs caught three formations of SBs (totalling 25 aircraft) from 54.SBAP returning from a raid via the Karelian Isthmus. Forty minutes later, nine bombers had met their end and several more were damaged – amongst the victorious pilots was 1Lt Urho Nieminen, who claimed two in FR-98.

On the 19th Nieminen, now in FR-78, and SSgt Virta, flying FR-84, each became aces when they downed an SB bomber apiece over the Karelian Isthmus. Following these reversals, Soviet bombers avoided the airspace over south-eastern Finland for almost two weeks.

However, in other areas along the massive frontline Soviet bombers continued to venture forth. Indeed, 24 hours after Nieminen and Virta had 'made ace', Detachment Luukkanen engaged SBs of 21.DBAP north of Lake Ladoga, attacking the bombers as they crossed the Finnish frontline, and then again as they departed the target. Five were destroyed, WO Viktor Pyötsiä (in FR-110) claiming two and SSgt Pentti Tilli (in FR-107) one – both duly became aces. However, their luck then swiftly changed as two I-16s (almost certainly from 49.IAP) belatedly appeared on the scene to chase the D.XXIs off, and Tilli was shot down and killed.

That same day 1Lt Tatu Huhanantti took off from Tampere in FR-91, which had just been repaired at the State Aircraft Factory. On his way back to base he encountered three SB bombers of 36.SBAP, quickly shooting down two of them before the five escorting I-153s could interfere. The wrecks were easily found as the aeroplanes had come down 60 kms north of Helsinki near a main railway line.

The excuse for a gathering on this occasion was that it was New Year's Day! Standing, from left to right, are WO Yrjö Turkka, Sgt Lasse Heikinaro, 1Lt Jorma Sarvanto and Danish volunteer 1Lt Erhard Frijs. Seated, from left to right, are Sgt Risto Heiramo, Sgt Eero Kinnunen and Sgt Tauno Kaarma. Visible in the distance is FR-112 'Black 7', which was usually flown by 1Lt 'Joppe' Karhunen (*J Sarvanto*)

3/LLv 24's FR-76 was so badly damaged in combat on 5 March 1940 that Sgt Mauno Fräntilä was forced to land between the Finnish and Soviet frontlines at Virolahti, on the north shore of the Gulf of Viipuri. The wounded pilot succeeded in making it back to friendly territory, whilst his aircraft became the target for Finnish mortar fire. Despite suffering further damage, LLv 24's oldest D.XXI (see the photograph on page 8) was successfully salvaged by the Red Army and put on display in Leningrad soon after the cessation of hostilities (*via P Manninen*)

Although up until now the Soviet fighters had shot down just two D.XXIs, there had been a number of close calls when pilots had attempted to engage the Polikarpovs despite Maj Magnusson's order forbidding such acts. In an effort to further discourage these dogfights from taking place, on 28 January *Lentorykmentti* 2 commander, Lt Col Lorentz, issued a ban on searching out and engaging enemy fighters for combat. Only bombers were to be attacked, for the D.XXI was no match for the more manoeuvrable I-153s and I-16s.

On 29 January Red Army artillery commenced shelling Finnish positions on the Karelian Isthmus with the aid of Polikarpov R-5 fire control aircraft. The local army commander duly contacted his counterpart in *Lentorykmentti* 2 and requested that fighters be scrambled to put a stop to the deadly accurate barrage. 1Lt Jorma Karhunen was given this task;

'At 1455 hrs I got an order to take off with three other Fokkers and head to Summa, where we were to either drive away or destroy two artillery fire control R-5s. Five minutes later the swarm sped across the ice and took off. My wingman, WO Yrjö Turkka, and I formed the lead pair, with 2Lt Olli Mustonen and Sgt Tauno Kaarma just astern of us.

'I had quickly devised a battleplan, which had to be foolproof, for we knew that the R-5s would be called off at the merest hint that we had been scrambled. We also knew that once we had left the frontline they would pop up again. I therefore decided to fool their ground-based spotters by flying along the west coast of Viipurinlahti just above the cloud line, which was at 2000 metres.

'We continued south, and at Koivisto we turned towards Summa. I then led the swarm into the clouds, before pulling my FR-80 abruptly out as we approached Summa – I needed to to make sure that the "patients" were still there. I soon spotted the R-5s as they lazily circled around relaying instructions to the artillery, although they were still some way away, so I quickly flew back into cloud. The next three minutes seemed to last forever as we closed on the enemy. Then we bounced them out of the clouds.

'One R-5 was conveniently below the lead pair as we burst out of the cloud, and 'Daddy' Turkka and I fired simultaneously at it. In seconds the aircraft's wings folded up as the fuselage burst into fire – the flaming ball of fire came down between the lines. Then we all fired at the second R-5, and it crashed behind the the Soviet frontline. Our job was done.

'We then managed to negotiate the enemy's ferocious anti-aircraft fire and return to base unscathed, which was a miracle in itself.'

Winter War survivor FR-116 of 5/LLv 24 is seen at Joroinen on 8 April 1940. Although barely visible on the original print of this photograph, there is a blue '4' painted on the fighter's rudder, denoting its allocation to the 5th Flight. Along with the 1st Flight, 5/LLv 24 was manned by pilots from LLv 26 until they returned to their original unit on 1 February 1940. And it was 'on loan' pilot 2Lt Kauko Linnamaa that had shared in the destruction of a DB-3M bomber on 21 December 1939 whilst flying FR-116

One of the Soviets' favourite targets during the Winter War was the State Aircraft Factory at Tampere. Indeed, it was so regularly bombed that a local defence flight was organised by its test pilots, and they were occasionally assisted by frontline pilots who found themselves caught up in the midst of a raid whilst picking up a repaired aircraft. 2Lt Olli Puhakka became embroiled in just such a situation on 29 January;

'Having just completed an interception, 1Lt Visapää and I were just about to land at Tampere when we observed a large twin-engined aircraft flying a transverse course to us. 1Lt Visapää was about 200 metres ahead of me, and almost directly above the runway when we spotted the bomber.

'My leader instantly fired at the aircraft, but I instead chose to close on it, leaving 1Lt Visapää behind me in a climbing turn. I was about 500-700 metres behind the enemy aeroplane, and as I gave chase I quickly realised that the distance between us was increasing rather than closing. In desperation I started to fire short bursts at the bomber, and the first one passed below it. However, the second cannon shell struck the aircraft near its port engine and the third burst hit the starboard wing.

'I assumed that I had damaged one of its engines, for I instantly overhauled the bomber, but now my guns refused to work! All I could do was to carry out some diving passes on the aeroplane, and I left it between Lempäälä and Viiala. I thought that 1Lt Visapää would now attack it, but he had lost the bomber in cloud at the same time as I broke off my attack.

'Luckily for me the bomber later crash-landed. I assumed that the entire action had been witnessed by 1Lts Itävuori and Visapää. Not knowing what had caused the aircraft to come down, I hoped that information from *T-LentoR 2* (*Täydennyslentorykmentti 2*, or Supplementary Flying Regiment 2, which was part of LentoR 2, in charge of advanced fighter combat training) would confirm whether its demise had been caused by cannon shells or machine gun bullets – the latter could also have been fired by 1Lt Visapää.

'I was flying FR-76, which was equipped with a 20 mm cannon under each wing, although only one had worked. I used all 18 rounds in the

drum. The fuselage machine guns had fired about 60-70 rounds when the synchronising gear broke down.'

The downed bomber was a DB-3M from 43.DBAP, its crew having been captured following a forced-landing at Urjala. Upon examining the wreckage, *T/Lento R2* did indeed find several cannon holes in the wings, which gave Puhakka credit for the victory.

On 30 January LLv 24's 2nd Flight suffered a major blow when its leader, 1Lt 'Jaska' Vuorela (flying FR-78), crashed to his death in poor weather at Ruokolahti. 1Lt Leo Ahola duly assumed command of the unit, forming Detachment Ahola, and leaving 1Lt Jorma Karhunen to head the remainder of the 2nd Flight. LLv 24 was now down to just 28 serviceable D.XXIs, and ten of its 'hired' pilots from LLv 26 returned to their original unit at month-end following the arrival of 30 ex-British Gloster Gladiator IIs. January's score for the D.XXIs was 34 aircraft shot down.

Seen on page 13 forming the backdrop for the group shot of pilots from 4/LLv 24, FR-110 looks a little worse for wear in this 8 April 1940 view. Its port ski had fallen off whilst the fighter was in flight, so its pilot, 3/LLv 24's 2Lt Olli Mustonen, had to effect an emergency recovery at Joroinen. The second-highest scoring fighter of the Winter War, FR-110 had been used by WO Viktor Pyötsiä to claim all 7.5 victories credited to him – the fighter was coded 'Blue 7' at the time. It is the only D.XXI to have been photographed bearing victory markings, these taking the form of four-and-a-half vertical bars on the starboard side of the fighter's fin (visible just forward of the '7' on the Fokker's rudder)

TROOP ESCORT

By late January the Soviet fighters committed to the Winter War had become very aggressive, flying in regiment strength (30 to 40 aeroplanes) deep into Finland. Newly-introduced drop tanks also meant that bombers could now be escorted on most raids, and the pilots of LLv 24 found it increasingly difficult to claim further kills.

Following a standstill along the frontline throughout the first weeks of 1940, on 1 February 1940 the Soviet Union launched the second phase of its offensive to capture the Karelian Isthmus. Other fronts remained static as the communists committed virtually all of their ground and air forces to achieving a breakthrough in this strategically important area. The inland

8 April 1940 proved to be something of a chequered day for LLv 24, as FR-117 was also damaged in a landing accident at Joroinen when the 1st Flight's MSgt Lauri Rautakorpi slid the fighter into a camouflaged barn! 'White 8' was normally assigned to 2Lt Olli Puhakka, who had been 'loaned' to LLv 24 from LLv 26. Finland's sixth highest scoring ace of World War 2, Puhakka went on to score 42 victories and win the Mannerheim Cross (on 21 December 1944)

bomber offensive also immediately switched to supporting the infantry, and large fighter formations started patrolling over the battlefield.

Fighting for its survival, the Finnish Army began a slow retreat a fortnight later, before finally halting the onslaught in front of Viipuri on 26 February. Throughout the offensive the D.XXI pilots concentrated on providing fighter protection for the troops in the frontline, and for replacements being sent into battle – LLv 24 flew up to 88 sorties a day.

With the increase in activity came further losses, and on 1 February I-16s from 7.IAP shot down and killed 2Lt Tapani Harmaja in FR-115. A similar fate befell Danish volunteer 1Lt Fritz Rasmussen the following day (in FR-81) when he was downed by I-16s from 25.IAP.

Within hours of the offensive being launched, Detachment Ahola flew to Turku, on the south-west coast, in order to protect this major port, and to prevent Soviet bombers from flying northward along the west coast.

On 3 February 1Lt 'Zamba' Sarvanto (in FR-80) claimed his tenth kill when he shot down a DB-3M of 51.DBAP at Nuijamaa, in south-eastern Finland. Later that same day, in the west, four D.XXIs bounced three DB-3s of 10.ABr (aviation brigade) and shot them all down into the Turku archipelago – 2Lt Pekka Kokko claimed two in FR-86. The following day another DB-3 was lost in the same area, and the Russians became alarmed, for they had not expected to find fighters so far west.

On 10 February the 4th Flight attacked a large bomber formation over Lappeenranta, but the equally numerous fighter escort of 7. and 25.IAP defended their charges well, shooting down MSgt Väinö Ikonen in FR-102 – he escaped with wounds. Big claims were now a thing of the past for the Fokker pilots, with the daily scores rarely bettering three bombers at most. Losses also continued to accumulate, and on 19 February a second Danish volunteer in 1Lt Erhard Frijs (in FR-80) was killed near Käkisalmi by I-16s of 25.IAP. One week later Sgt Tauno Kaarma escaped with injuries when I-16s of 68.OIAP downed his FR-85 at Immola.

February ended badly for the Finnish fighter arm when, on the 29th, Soviet fighters carried out a series of air raids on the airfields occupied by LLv 24 and 26. 49.IAP claimed a Gladiator of LLv 26 (assigned to

5/LLv 24 D.XXI FR-105 'White 5' is seen at Joroinen in April 1940. It was assigned to Sgt Eero Kinnunen at the time this photograph was taken, although he claimed all of his 3.5 Winter War kills in FR-109. FR-105 did enjoy success, however, whilst being flown by future aces Sgts Lasse Aaltonen and Onni Paronen (then both members of LLv 26). The fitment of skis in place of wheels on the D.XXI's fixed undercarriage legs had very little effect on the fighter's performance. On 19 April 1940 LLv 24 exchanged its surviving Fokker fighters for the Brewster Model 239s flown by LLv 32 (ex-LLv 22)

LLv 24's Detachment Luukkanen, which had moved to Ruokolahti three weeks earlier) during a morning sweep, and then at noon a 'bomber' formation was detected approaching Ruokolahti. These aircraft turned out to be six I-153 'Chaikas' and 18 I-16 'Ratas' from 68.OIAP (O, in this instance, standing for detached), which took the Gladiator pilots by complete surprise as they took off. Three were immediately destroyed, and in the low-altitude combat which ensued, two more were lost, along with the D.XXI (FR-94) of 1Lt Tatu Huhanantti – all six pilots perished. Only one I-16 was downed in return, whilst another clipped some trees and crashed.

Despite increased opposition, LLv 24 had claimed 27 aircraft destroyed in February, although it was now down to just 22 serviceable D.XXIs.

FINAL ROUNDS

On 1 March 1940 unit moved its headquarters to Lemi, and within two days all five flights had followed suit.

Following the bitter fighting of February, the Finnish Army had finally withdrawn from the Karelian Isthmus by 1 March, although it remained steadfast in front of Viipuri. Sensing retreat, the Red Army started crossing the frozen Gulf of Finland west of Viipuri the following day.

Soviet forces had soon established two small bridgeheads on the mainland, and in order to prevent a full-scale invasion, the entire Finnish Air Force was sent into action against troops, tanks and supply columns crossing the ice. The aircrews went about their task with clinical precision, light and medium bombers stopping the motorised units in their tracks and fighters strafing the infantry. The Red Army was caught totally exposed on the vast open spaces of the icefield, and within a week the invasion had been suppressed. This intensive operation also attracted Soviet fighters into the area, and Capt 'Eikka' Luukkanen, who lead several strafing missions during this time, describes one such sortie flown on 5 March;

'I broke radio silence and ordered the formation (of 15 Fokkers) out of the clouds. In an effort to confuse the enemy, we approached them from their direction. I banked to the left and led the formation into a dive. There seemed to be no lack of targets, with columns of cars, trucks and tanks filling the ice for the full four kilometres that separated Tuppura from Vilaniemi. A squadron of I-16s was spotted circling above Uuras, and another fighter unit was observed on the other side of the gulf over Ristiniemi. I continued my shallow dive, knowing that the sooner we hit

Although now assigned to LLv 32, D.XXI FR-108 is seen here running up at Siikakangas in late May 1940 whilst still wearing the colours of its previous user, 3/LLv 24. 3rd Flight leader Capt Eino Luukkanen had flown this aircraft ('blue and white 6') during the Winter War, claiming one Soviet aircraft solely destroyed and another shared

Photographed between sorties at Siikakangas in June 1940, FR-92 was the mount of 4/LLv 24's deputy leader, 1Lt Per-Erik Sovelius. One of ten Winter War aces, he claimed all six of his victories with FR-92. The fighter still sports the tactical marking 'Black 5' favoured by 'Pelle' Sovelius

the targets on the ground the better. We were now only one kilometre away from the nearest vehicles, and the sky all around us was filled with explosions and tracers of anti-aircraft fire. Both white and black clouds of flak exploded nearby, immediately attracting enemy fighters.

'My first burst of fire hit an infantry column, and next in my gunsight loomed a line of trucks, followed by two tanks. The bullets from our rifle-calibre guns did not seem to have any effect on the latter, with the tracer rounds visibly bouncing off their armour plating. Following my strafing run I looked back to see if the rest of the formation had followed my lead.

'Immediately after the attack we flew due west at low-level towards friendly territory, before banking to the north heading for our base. By following such a course, we prevented any aircraft that happened to be tailing us from finding our base.'

LLv 24 flew 154 assault sorties during the seven-day battle on the ice, claiming one I-16 shot down and losing Sgt Mauno Fräntilä (in FR-76) to fighters from 7.IAP fighters. Fräntilä, who had been on the strafing mission described by 'Eikka' Luukkanen, was wounded during the course of the mission and forced to crash-land between the lines at Virolahti. He managed to reach the Finnish frontline, but his badly damaged D.XXI was seized by the Red Army and displayed in Leningrad as a war trophy.

The situation on the Karelian Isthmus was still critical when peace negotiations commenced in Moscow on 8 March. The resistance of the Finnish forces, supported by material help from western nations who

Yet another ex-LLv 24 machine, FR-95 was photographed at Siikakangas in June 1940 whilst serving with LLv 32. This aircraft had been damaged in a landing accident early on in the Winter War, and after being repaired, it was issued to 4/LLv 24's MSgt Lasse Heikinaro on 3 February 1940. He went on to claim three victories with it prior to the ceasefire coming into effect. 'Black 6' remains painted in the colours it wore during the recent conflict. In September 1940 LLv 32 at last introduced its own tactical numbering system (*U Nurmi*)

Well-known Finnish company Nokia Oy donated sufficient funds for the air force to purchase Brewster Model 239 BW-355. In return the aircraft was adorned with the inscription *NOKA*. One of the first examples of the American fighter to arrive in Finland, it is seen at Hollola in late March 1940 whilst assigned to LLv 22. On 18 April BW-355 was transferred to LLv 24, and the aeroplane continued to carry this name through to its destruction on 24 October 1944

threatened to join in militarily if the invasion continued, convinced the USSR that further action would only see the war expand into an international crisis, which they did not want. So, on 13 March at 1100 hrs a cease-fire commenced. Accordingly, but unjustifiably, Finland handed over those tracts of land that the Soviets had demanded back in late 1939.

When the Winter War commenced LLv 24 had 35 serviceable D.XXIs on strength, and by the time it came to an end this number had been reduced to 22. The unit had flown 2388 sorties and claimed 120 aircraft shot down (100 of these were bombers). It had lost 11 aircraft to all causes – nine in combat, one to 'friendly' flak and one in a flying accident. Seven pilots had also been killed. These figures also include the claims made and the losses suffered by the D.XXI pilots 'on loan' from LLv 26.

The Fokker fighter had proven itself to be a very reliable interceptor, LLv 24's groundcrewmen succeeding in keeping aircraft airworthy in the most severe winter weather thanks primarily to their simplicity of construction. The D.XXI bore the brunt of the fighter operations, for the Gladiators only became operational on 2 February (claiming 37 aerial victories), the Morane-Saulnier MS.406s on 17 February (accumulating 14 kills) and the Fiat G.50s on 26 February (claiming 11 shot down).

The Finnish Air Force claimed 207 aircraft shot down and anti-aircraft artillery batteries were credited with a further 314, resulting in an overall total of 521 destroyed. Contemporary Soviet records stated that 261 aircraft were lost, although recent figures sourced from Russian archives show that 579 were actually shot down – 58 more than the Finns claimed!

FOKKER D.XXI ACES

Rank	Name	Flight(s)	Victories
1Lt	Sarvanto, Jorma	4, 1	13
WO	Pyötsiä, Viktor	3	7.5
1Lt	Huhanantti, Tatu+	3	6
MSgt	Virta, Kelpo	2	6
1Lt	Sovelius, Per-Erik	4	5.5
SSgt	Tilli, Pentti+*	3	5
1Lt	Nieminen, Urho*	5	5

+ – killed in action
* – member of LLv 26

FINNISH OFFENSIVE

On 9 April 1940, just four weeks after the end of the Winter War, Germany attacked Denmark and Norway. Within days of the invasion being launched, the *Wehrmacht* had occupied most of Scandinavia, bar neutral Sweden and Finland. The following August the Soviet Union annexed Estonia, Latvia and Lithuania, leaving Finland almost totally isolated geo-politically. The USSR bordered it to the east, neutral Sweden to the west and Germany to the south. The latter nation was no ally of the Finns at the time, having offered no assistance during the Winter War – the Nazi regime had actually obstructed the supply of military equipment during the conflict.

On 18 March 1940 LLv 24 moved to Joroinen, where it was re-organised into just four flights. The following month, on 13 April, the unit sent 16 of its D.XXIs to the State Aircraft Factory for their skis to be replaced by wheels, and for much-needed maintenance to take place. Two of the pilots were delayed in their departure from Joroinen, and while attempting to catch up with their squadronmates, 2Lt Eero Savonen (in FR-93) and 2Lt Heikki Ilveskorpi (in FR-101) collided in mid-air and were killed.

The unit remained at Joroinen for just another six days after bidding farewell to its Fokker fighters, moving to Helsinki-Malmi on 19 April. Here, the unit was issued with the Brewster Model 239s that had briefly served with LLv 32 (formerly LLv 22). Although this swap was not popular with the personnel of the latter unit, the men of LLv 24 had clearly proven themselves to be Finland's crack pilots during the Winter War, and they deserved the best aircraft that the air force had to offer.

Having completed its conversion onto the Brewster at Helsinki-Malmi during the early summer of 1940, LLv 24 moved to the newly-constructed airfield at Vesivehmaa in August. Its former base had been just ten kilometres from the city centre, and had not proven particularly suitable for the intensive flying training that the unit was now undertaking. Vesivehmaa, however, was 80 km north of the Finnish capital, and isolated from any major conurbation.

In the 14 months of peace from March 1940 through to June 1941, LLv 24 worked hard to train all of its regular and reserve pilots to be ready for a second war with the USSR. During this period, two pilots were killed in accidents – Sgt Risto Heiramo crashed into the railyard at Toijala in BW-360 in bad weather on 14 October, and WO Kelpo Virta struck the ground at Vesivehmaa in BW-391 during a demonstration flight on 28 January 1941.

THE CONTINUATION WAR

The German surprise attack on the USSR, codenamed Operation *Barbarossa*, was revealed to Finnish military leaders four weeks prior to it commencing on 22 June 1941. Armed with this knowledge, the Finns instigated a full-scale mobilisation four days before the invasion.

Soon after the start of *Barbarossa*, Soviet intelligence discovered a large number of German aircraft based on Finnish airfields, which made the communists fearful of a massive air raid being launched on Leningrad from this direction. On the Finnish front, which stretched from the Gulf of Finland to the Arctic Sea, the Red Air Forces were equipped with 224 fighters and 263 bombers, and early on the morning of 25 June 1941, about 150 of the latter took off and attacked several locations in southern Finland. So began the Continuation War.

The Finnish fighter force was now in better shape than it had been some 19 months earlier, and *Lentorykmentti* 2 commander, Lt Col Richard, committed three full-strength squadrons to the defence of the country. These units were LLv 26 with Fiat G.50s, LLv 28 flying Morane-Saulnier MS.406s and LLv 24 with its Brewsters, split into the following flights;

***Lentolaivue* 24 on 25 June 1941**

Commander	Maj Gustaf Magnusson with HQ at Vesivehmaa
1st Flight	Capt Eino Luukkanen at Vesivehmaa with nine Brewsters
2nd Flight	Capt Leo Ahola at Selänpää with eight Brewsters
3rd Flight	1Lt Jorma Karhunen at Vesivehmaa with eight Brewsters
4th Flight	1Lt Per-Erik Sovelius at Vesivehmaa with eight Brewsters

BREWSTER MODEL 239

Following the outbreak of the Winter War, the need for Finland to rapidly purchase fighter aircraft led to a deal being struck with the US government on 16 December 1939. The Brewster Aeronautical Corporation would sell 44 Model 239s earmarked for the US Navy to Finland at 2.7 million Finn marks per aircraft, and in turn keep its primary customer happy by replacing these fighters with the improved F2A-2. All naval equipment was removed from the airframes, and their 940 hp Wright R-1820-34 Cyclone engines replaced by a civil version of the Wright Cyclone (the 950 hp R-1820-G5).

Thirty-eight aircraft came from the US Navy's F2A-1 block (construction numbers 18 to 55) and six from the Belgian Model 339B block (construction numbers 57-62), all of which were standardised as Model 239s. The airframes and engines were packed into crates and, from 13 January 1940 onwards, shipped to Stavanger, in Norway. From here they continued their journey by train to the SAAB plant at Trollhättan, in Sweden, where assembly took place. The aircraft were given serial numbers BW-351 to BW-394 inclusive, and during the transfer flight to Finland all markings were covered.

The first four Brewsters – with Finnish fighter pilots at the controls – were flown into Finland on 1 March 1940, and they were followed by two more one week later. These were the only examples to actually arrive during the Winter War, with the last of the 44 Model 239s being flown in on 1 May 1940.

Brewster Model 239 BW-356 was captured on film by the air force depot photographer at Tampere on 29 May 1940. Newly assigned to LLv 24, it is still painted in the factory-applied aluminium dope finish worn by all the Brewster fighters delivered to Finland in 1940. These aircraft would remain in this scheme until 19 June 1941, when the unit camouflaged all of its Model 239s with black and olive green upper surfaces (white replaced much of the black in winter) and light grey undersurfaces (*Finnish Air Force*)

Mechanic Ahti Askola helps Lt Col Gustaf Magnusson strap into Brewster BW-380 at Hirvas on 28 June 1942. 'Eka' Magnusson led LLv 24 until the end of May 1943, when he was appointed commander of *Lentorykmentti 3*, which controlled three fighter squadrons. Considered the 'Father' of the modern Finnish fighter force, Magnusson flew 158 sorties and destroyed 5.5 bombers during his time in the frontline. He was duly awarded the Mannerheim Cross on 26 June 1944 for both his outstanding leadership qualities and the creation of an indigenious fighter command and control system (*SA-kuva*)

DAY ONE

The large bomber formations heading for southern Finland were first spotted at 0700 hrs, and the news was quickly relayed to Selänpää, where Maj Magnusson had forward deployed 2/LLv 24 in anticipation of just such an air raid. At 0710 hrs two Brewsters were scrambled, with SSgt Kinnunen flying BW-352 and Cpl Lampi BW-354. In the following account, Kinnunen describes the first aerial engagement of the Continuation War;

'We took off after receiving a message that Inkeroinen was being bombed. Once airborne, I observed a 27-bomber formation heading north-west. Although I quickly caught the bombers up, Cpl Lampi had beaten me to them and already shot one down in flames.

'I chose my own target, and this too caught fire following a couple of short bursts. I then sent a second bomber down in flames and fired at three more, all of which seemed to start smoking. My fighter was then hit by return fire, and having received a shrapnel wound to my hand, I was forced to break off my attack and return back to base. I approached each of the bombers that I attacked either directly from behind, or from below and behind.

'The bombers all had ventral guns, and my aeroplane suffered three hits – one in the cockpit and two in the rudder.'

The Finnish pilots had engaged 27 SBs from 201.SBAP at a height of 1500 metres as they approached Heinola, the Soviets losing five bombers in total. Kinnunen and Lampi were both credited with destroying 2.5 aircraft apiece, with the former becoming an ace on this sortie, for he had previously scored 3.5 kills during the Winter War.

Further interceptions later that morning saw the Brewster pilots destroy five more SBs, with WO Turkka (in BW-351) downing two to add to his score of 4.5 from the Winter War. SSgt Kinnunen also claimed two more during his second sortie of the day, raising his tally to 4.5.

Maj Magnusson's BW-380 is manhandled onto a petrol drum in preparation for a spell in the firing butts at Rantasalmi on 10 July 1941. Magnusson favoured 'Black 1' as his tactical number, here applied in the colours of the unit's 4th Flight. Three days prior to this photograph being taken, Magnusson had scored his last kill – a DB-3M bomber – in this very machine. None of his contemporaries (squadron commanders) within the Finnish fighter force achieved comparable success (*SA-kuva*)

Recent research has shown that ten SB bombers (three from 2.SBAP, six from 201.SBAP and one from 202.SBAP) were lost in LLv 24's operational area that day, which matches the figure claimed by the Brewster pilots.

A haul of 26 bombers destroyed (23 now acknowledged) was just the start the Finnish fighter force wanted to the Continuation War, although its ground-based early warning and fighter control system proved to be less than efficient – indeed, despite having 125 fighters on duty, less than a fifth engaged the enemy. This problem was slowly put right, however.

For the remainder of the month LLv 24 flew combat air patrols over the southern coastline in an effort to prevent Soviet bombers from entering Finnish airspace – this was a difficult task, for the frontline stretched from Leningrad to Haapsalu, in eastern Estonia. During one such patrol, on 29 June, 1Lt Pekka Kokko (in BW-379) encountered two Beriev MBR-2 flying boats and shot them both down into the sea. These kills gave him

Gun tests completed, BW-380 has had its cockpit, engine and tail covered with tarpaulins at Rantasalmi to keep the damp out overnight – LLv 24's aircraft rarely enjoyed the luxury of sheltered accommodation. The unit's 'Lynx' badge is clearly visible on the fuselage, this marking having originally been applied to 3rd Flight Brewsters only. It took exactly three of the fuel drums visible in the foreground to fill the Model 239's two 300-litre integral tanks, which were located in the wings (*Bundesarchiv*)

BW-352 'White 2' of 2/LLv 24 basks in the evening sun at Selänpää on the opening day of the Continuation War, 25 June 1941. Earlier in the day, SSgt Eero Kinnunen had downed four Tupolev SBs, and shared in the destruction of a fifth, during the course of two sorties. The flight claimed ten bombers in total over southern Finland on the 25th, all of which have been confirmed by recent research (*SA-kuva*)

2Lt Matti Pastinen was LLv 24's sole casualty for the first six months of the Continuation War, the 3rd Flight pilot trying to take off from Vesivehmaa on 28 June 1941 with the propeller of BW-369 'Orange 7' set in coarse pitch. Failing to clear the trees at the end of the runway, he cartwheeled several times before his fighter came to rest virtually split in two. Extricated from the wreckage with terrible injuries, Pastinen died in a field hospital five days later (*R Lampelto*)

ace status, following on from his 3.5 victories gained during the Winter War. Kokko's combat report recounted;

'I took off at 1040 hrs on a combat air patrol from Kotka to Porvoo, with Cpl Mellin on my wing. At 1135 hrs I observed two flying boats heading for the Finnish coast from Suursaari, and six minutes later we intercepted them. My first attack was carried out alone because Cpl Mellin had lagged too far behind me in the final pursuit of the enemy.

'I dived at the trailing flying boat, but only succeeded in causing slight damage – I observed a faint trail of smoke – due to my fighter's small calibre fuselage gun being the only weapon that was operable. I then made a second dive towards the lead aeroplane, and this time my heavy fuselage gun worked as well, and the flying boat instantly burst into flames after I fired at its engine.

'In the meantime Cpl Mellin had attacked the slightly damaged flying boat, but had failed to shoot it down. I went after it for a second time, putting a short burst of fire into the engine, which was soon ablaze. Both aeroplanes crashed into the sea.

'The flying boats were slow and clumsy, with gunners in the nose and above the rear fuselage. My aeroplane suffered no damage.'

That same day Finnish pilots reported their first encounter with the Red Air Force's newest bomber, the Petlyakov Pe-2. Only available in small numbers during the first months of the Continuation War, early Pe-2s were used primarily as photo-reconnaissance aircraft. Ranking Winter War ace 1Lt Jorma Sarvanto was amongst the first Finnish pilots to engage the Pe-2, intercepting a pair of bombers during a combat air patrol flown in BW-357 on 29 June;

'Two twin-engined, twin-tailed aeroplanes overflew Kuusankoski at 3000 metres, heading towards Utti. They were separated by a distance of 600 metres. I radioed my wingman to attack the bomber to the south of Kuusankoski while I went after the one to the north. Whilst pursuing my quarry, "friendly" anti-aircraft tracers and explosions surrounded me.

'The enemy was flying at about the same speed as I was, and I aimed carefully and fired a burst at a distance of 400 metres, hitting the port engine and causing it to emit smoke. I closed on the bomber and fired again from behind and below. It crashed between Sippola and Utti after shedding some fuselage plating. One man bailed out.

'I then chased after the second bomber at full speed as far as the coast at Seiskari, but it continued to pull away and eventually escaped.

'The downed machine had opened fire at a distance of about 800 metres when I turned after it. I was doing over 500 km/h.'

Contemporary Soviet records show that during the bomber offensive that lasted from 25 June to 1 July, 39 Finnish and German (there were Luftwaffe detachments on two airfields) air bases were attacked and 130 aircraft destroyed on the ground. The German records show no such losses, whilst the Finns had just two aircraft lightly damaged. On the other hand, Finnish fighters claimed 34 bombers destroyed during the same period – LLv 24 was credited with half of these.

The only squadron loss to occur at this time claimed the life of 2Lt Matti Pastinen, who crashed in BW-369 on 28 June at Vesivehmaa when his fighter failed to take-off due to the pilot having chosen the wrong propeller pitch setting. Pastinen died of his injuries five days later.

KARELIA OFFENSIVE

Following the early success of *Barbarossa*, the Finns hastily planned an offensive of their own in late June 1941. It consisted of a two-phase assault

BW-372 'White 5' was the mount of 2nd Flight boss, Capt Leo Ahola, and it is seen at Joensuu in July 1941 – his flight flew bomber escort missions from this base during the last week of the month. Following the Luftwaffe's adoption of yellow eastern front theatre markings, all Finnish aircraft were painted with a 50-cm yellow fuselage band and a similarly-coloured wing tip flash which usually covered a sixth of the total span (*E Rinne*)

BW-368 'Orange 1' was assigned to 3/LLv 24 in July 1941, and it was photographed 'under wraps' at Rantasalmi soon after the commencement of the Continuation War. Sgt 'Nipa' Katajainen became an ace in this aircraft on 12 August 1941 when he downed two I-153s – he would claim 8.5 victories with BW-368 between 1 August 1941 and 23 September 1943. Katajainen joined LLv 24 as a young corporal upon its mobilisation on 18 June 1941. Note BW-368's distinctive unit badge, which was approved as a result of air force C-in-C, Maj Gen Lundquist, permitting such personalisation of aircraft just one week after the mobilisation of the air force. No such unit or individual emblems had previously been allowed on military aircraft. Only a handful of outfits took advantage of this relaxing in the regulations

on Karelia, north of Lake Ladoga, and a single attack across the Karelian Isthmus. The army's objective was to seize only those areas ceded to the USSR as part of the 1940 peace treaty.

A Karelian Army was formed for operations north of Lake Ladoga, and it was assigned *Lentorykmentti* 2, with all three fighter squadrons, plus LLv 12 and 16 for reconnaissance and army co-operation duties. *Lentorykmentti 4's* Bristol Blenheim bombers could also be called upon when needed.

On 3 July most of LLv 24 flew from Vesivehmaa to Rantasalmi, in eastern Finland, so as to be closer to the forthcoming operational area.

The massing of Finnish troops in preparation for the offensive failed to escape detection by Soviet reconnaissance aircraft, however, and on 8 July air attacks commenced. Defending the assembled force, Brewster pilots shot down two bombers and six fighters during the course of three engagements. One of those pilots to enjoy success on this day was Winter War veteran MSgt 'Lapra' Nissinen, who was seeing his first action of the new conflict. Assigned to the 3rd Flight, he was airborne in BW-353 over the front at 1400 hrs when he spotted enemy aircraft;

SSgt Jalo Dahl poses in front of 4/LLv 24's BW-393 'Black 9' at Rantasalmi in August 1941. The flight colours worn on the fighter's rudder were basically the same as had been used by the unit in the Winter War, with the addition of an identification stripe on the spinner

'I was flying in 1Lt Kokko's swarm, leading the top pair, when I observed two "Chaikas" and an R-5 over Enso at an altitude of about 500 metres below and to the left of me. I dived down on the R-5, holding my fire until I was 100 metres away from it. Tracer rounds hit the fuselage behind the pilot, and I noticed that the gunner did not shoot back at me.

'I then commenced a turning fight with a "Chaika", and was twice forced to break off by diving, after which I pulled steeply up and attacked my opponent head-on. We came towards each other on several occasions, and both of us were shooting at each other. On our final pass, I succeeded in hitting the "Chaika's" engine, which belched out smoke and the gear dropped down.

'I could not follow the aeroplane down, however, for another "Chaika" had now slipped in behind me. After losing him in a dive, I then turned back towards my opponent and attacked him from head-on too. After flying at each other a couple of times, the "Chaika" pilot tried to break off the fight and escape by diving to the ground. I easily caught him up, and he apparently failed to notice that I was behind him for he kept on flying straight ahead. By this stage only my starboard fuselage gun was working, so I opened fire close behind the "Chaika's" tail. The fighter immediately caught fire and crashed in the woods at Enso on the west side of the river. I was only 20 metres above the ground when I opened fire.

'The "Chaika" was more manoeuvrable than my Brewster, and after four or five tight turns it was able to get onto my tail. My fighter was considerably faster than the Russian aircraft, however.'

The standard camouflage pattern for Finnish fighters is clearly shown in this tree-top view of BW-367 'White 7', taken at Rantasalmi in early August 1941. Assigned to 2/LLv 24, the Brewster was undergoing an engine overhaul when this photograph was taken. Future 5.333-kill ace WO Veikko Rimminen had used this aircraft to down an SB-2bis on 30 June 1941. The unit's 'striking lynx' motif had initially appeared on 3rd Flight Brewsters in early July 1941 – many of its personnel came from the Häme region of Finland, where the lynx was the 'county symbol' depicted on its coat of arms. Within two weeks the squadron's remaining three flights (including 2/LLv 24) had also adopted this emblem, and it duly became LLv 24's unit badge – not all Brewsters featured the marking, however. The lynx was applied in white through the use of a stencil. The badge remained on most Model 239s for over two years, and only disappeared in late 1943 when these aircraft were overhauled (and repainted) at Tampere

Having already donned their parachutes, four Winter War veterans of 3/LLv 24 run through a quick pre-take-off brief at Rantasalmi on 10 July 1941. They are, from left to right, WO Ilmari Juutilainen (flight leader), 1Lts Jorma Karhunen and Pekka Kokko and MSgt Lauri Nissinen. Their accumulated score by the end of World War 2 would total a staggering 171 aerial victories, and all bar Kokko would receive the Mannerheim Cross while serving with LLv 24 (*SA-kuva*)

Both of Nissinen's I-153s hailed from 65.ShAP, and they were seen to crash from Finnish ground observation posts. Having scored four kills during the Winter War, Nissinen's hard fought victories of 8 July made him an ace, and also resulted in him being warned against head-on attacks by his flight leader.

LLv 24 enjoyed even greater success the following day. Led by Maj 'Eka' Magnusson, 12 Brewsters of the 3rd and 4th Flights took off on a combat patrol at 0400 hrs, and 70 minutes later nine of the fighters engaged 15 I-153s of 65.ShAP over Lahdenpohja. 1Lt 'Pelle' Sovelius, piloting BW-378, was a participant in the combat that ensued;

'The fighting had already started when I spotted five I-153s in a loose formation 500 metres in front of me. I attacked one, which evaded, but another drifted across in front of me in a climbing turn so I shot at it. The covers behind the aircraft's engine immediately blew off and it crashed vertically into the ground, trailing a thick plume of smoke.

'I then climbed back up into the fight and hit two more I-153s with a series of well aimed bursts, before being forced to break off my attack after a further three "Chaikas" latched on to my tail.

'The I-153s stayed at the same altitude throughout the fight, and their only defensive manoeuvre was to bank downwards.'

In ten minutes the Red Air Force had seen eight of its 'Chaikas' destroyed and another four damaged, these being credited to six pilots – WO Juutilainen (in BW-364) and MSgt Nissinen (in BW-353) claimed two apiece, while 3rd Flight leader 1Lt Karhunen and 4th Flight leader 1Lt Sovelius fought with such tenacity that Maj Magnusson proposed to Lt Col Lorentz that both men be awarded the Mannerheim Cross. This honour was denied for the time being, however.

Flying an aircraft that was heavier than most of its rivals, which allowed the Brewster to attain greater speeds in a dive, Finnish pilots had been employing 'pendulum' tactics since the start of the Continuation War. Such tactics called for the pilot to dive at the enemy at great speed, make one firing pass, and then climb back up to altitude, before repeating the

BW-366 'Orange 6' is manually re-fuelled at Rantasalmi in July 1941 by a bronzed mechanic – note that he is wearing a knotted handkerchief on his head! This fighter was assigned to 3/LLv 24's flight leader, 1Lt 'Joppe' Karhunen, who, on 4 July 1941, claimed his first Brewster kill with BW-366. He continued to enjoy success with this particular machine through to 4 May 1943, when he claimed his 31st, and last, kill. No fewer than 17.5 of Karhunen's victories were claimed with BW-366, and a fair percentage of his 350 operational sorties were also flown in the fighter. Note that the aircraft's spinner was also painted orange to match its flight number (*R Lampelto*)

manoeuvre all over again. These tactics had been effectively employed by the Luftwaffe since the Spanish Civil War, and they would also bring Finnish fighter pilots great success for the next two years.

On 10 July 1941 the Karelian Army commenced its offensive, and within six days it had seized the northern tip of Lake Ladoga. Continuing to advance along the north coast, the infantry achieved their intermediary goal exactly two weeks later when they reached the River Tuulos. At this point the supreme commander of the Finnish armed forces, Marshal Carl Gustaf Emil Mannerheim, called a halt to the advance.

The offensive to capture the Karelian Isthmus started on 31 July, with the army spearhead pushing east past Viipuri and arriving on the shores of Lake Ladoga exactly one week later – it eventually met up with the victorious Karelian Army on 15 August. Viipuri was left in a state of siege until finally taken by Finnish troops on 30 August 1941. For four days

During the Continuation War LLv 24's general headquarters was located in the small town of Mikkeli, and four Brewsters from the 1st Flight were based at a nearby airfield for two months from 2 July 1941 in order to protect the HQ. 1Lt Joel Savonen's BW-361 'White 8' was one such aircraft, the fighter seen here sporting one victory bar gained on 16 July 1941. This was the first of Savonen's eight kills (*Bundesarchiv*)

following its capture, troops chased the Red Army in retreat eastward to the old Russo-Finnish border, where they were ordered to stop 30 km from Leningrad.

LLv 32, equipped with Hurricanes and Hawks (and reinforced by the Brewsters of 3/LLv 24, which flew top cover), flew close support sorties for the army throughout the campaign. The first major air battle to be fought took place on 1 August, when the Brewster pilots downed six I-16s north-east of Viipuri – WO Juutilainen claimed a double victory in BW-353.

Eleven days later, in the same area, Capt 'Joppe' Karhunen's six Brewsters succeeded in preventing more than 20 I-153s of the air forces of the 23rd Army from strafing Finnish infantry. The action started at 1300 hrs and lasted for a full 30 minutes, during which time nine 'Chaikas' were shot down. Every Finnish pilot was credited with at least one kill, whilst Sgt Nils Katajainen claimed two in BW-368 and WO Ilmari Juutilainen three in BW-364. The latter pilot remembers;

'As soon as 1Lt Strömberg shouted in the radio "'Chaikas' below, northbound", I spotted them. I repeated his call and then led Sgt Huotari in to attack the rearmost aircraft in the formation. Our squadronmates joined the fight a short while later. I counted 22 enemy aircraft prior to engaging them. I managed to take the Russian pilots by surprise, and after

4/LLv 24's deputy leader, 1Lt Iikka Törrönen (right), and his mechanic pose together in front of BW-385 'Black 2' at Rantasalmi in late July 1941. On 3 December 1941 this particular machine became the first Brewster to be lost as a direct result of enemy action when it was downed by Soviet anti-aircraft fire at Novinka. Its pilot, 1Lt Henrik Elfving, was killed

Cpl Paavo Mellin sits strapped into the cockpit of 3/LLv 24's BW-355 'Orange 7' at Rantasalmi in July 1941. Purchased for the air force by Nokia Oy (see the photograph on page 27), the aircraft's presentation inscription was reapplied in white paint when the fighter was camouflaged in June 1941. Future 5.5-kill ace Paavo Mellin shot down an I-16 (on 1 August 1941) and shared in the destruction of a MiG-3 (on 6 July 1941) whilst flying BW-355

On 26 September 1941 3/LLv 24's Sgt Nils Katajainen claimed his sixth kill (an I-153). Upon the ace's return to the flight's Mantsi base, on the dry shore of north-eastern Lake Ladoga, one of his mechanics hastily added another 'Chaika' silhouette (seen from head-on) to the bottom of his unique scoreboard, worn on the fin of BW-368 'Orange 1'. The remaining symbols to the right of Katajainen's shoulder denote three I-153s, an I-16 and an SB-2bis destroyed – all Continuation War victories. Katajainen continued to fly BW-368 until September 1942, when he was posted away from LLv 24 to learn to fly twin-engined aircraft (*N Katajainen*)

4/LLv 24's flight leader, and Winter War ace, Capt 'Pelle' Sovelius, poses at the dry shore landing strip at Lunkula in early October 1941. Behind him is BW-378 'Black 5', which was inscribed *OTTO WREDE* in honour of Swedish Baron Hugo Hamilton and his wife, who had donated funds for its purchase – one of Mrs Hamilton's relatives had been killed fighting with the Swedish voluntary battalion in Lapland during the Winter War. By the time 'Pelle' Sovelius had left the squadron on 16 February 1942, he had scored 12.5 kills (three and four shared destroyed in D.XXI FR 92 and seven in BW-378) during the course of 257 sorties

a long burst the first camouflaged I-153 that I fired at started to smoke, banked gently to the right and went vertically down. The pilot did not take to his parachute.

'The second one I attacked from above and behind, and it lost several panels whilst continuing to fly straight ahead, before entering into a spiral dive. I kept it in my sights for as long as I could, and again nobody jumped.

'The third I-153 that I attacked was again approached from the rear, and my Brewster was covered in its oil following my firing pass. The pilot failed to take any evasive action, and his fighter went into a dive, banking away to the right. I followed it down to a height of about 1000 metres, and it remained in this attitude until it hit the ground between Kirvu and Koljola. Once again, the pilot did not bail out.

'The fourth "Chaika" that I shot at was also attacked from astern, and it too started a slow banking turn. I fired one more burst into it before I dived past the I-153, and after climbing back up, I did not see it anywhere.

'I fired at ten aeroplanes in total, and they were all camouflaged.'

OCCUPATION OF OLONETS AND KARELIA

On 3 September 1941 the Karelian Army commenced its four-day, 75-km advance on the River Svir. The very next day, on 8 September, the German *Wehrmacht* reached Lake Ladoga in the south, and Leningrad's 900-day siege began. The Finns, meanwhile, continued their march both

Cadet Urho Sarjamo (centre) of 4/LLv 24 is seen with his mechanics near the tail of BW-380. On 18 August 1941 Sarjamo engaged the enemy for the first time, and whilst attempting to get on the tail of one of the eight Russian fighters that swarmed around him, he accidently closed BW-380's throttle, cutting all power to its engine. He successfully belly-landed the Brewster at Vuoksenranta, although it suffered extensive damage as it ground to a halt. BW-380 was duly sent to the State Aircraft Factory, which took exactly four months to effect repairs. BW-380 was eventually lost on 2 May 1943 when it was shot down by a LaGG-3 over Oranienbaum – the Brewster's pilot, 2nd Flight leader Capt Iikka Törrönen, was killed. 'Urkki' Sarjamo, meanwhile, went on to claim 12.5 kills prior to his death in combat on 17 June 1944 (*U Sarjamo*)

eastward and northward to Petrozavodsk, with LLv 24, 26 and 28 flying top cover for the advancing Karelian Army.

On the opening day of the offensive, Capt Luukkanen's strengthened 1st Flight flew to the temporary airstrip at Lunkula, which comprised a dry, sandy runway on the shore of Lake Ladoga's north-east coast. Maj Magnusson followed a fortnight later with the rest of the unit, while the 1st Flight moved still further south-east to Nurmoila, east of Lake Ladoga.

Up to this point in the Continuation War LLv 24's 2nd Flight had been exclusively performing the bomber escort mission, and on 6 November it relocated to Helsinki-Malmi to protect the capital from any possible bombing raids by the Red Air Force. To date, its pilots had had very little opportunity to claim any aerial victories.

This had not been the case for the 4th Flight, however, and on 17 September eight of its Brewsters, led by flight leader Capt Sovelius, jumped 14 MiG-3s east of Petrozavodsk. Exploiting their tactical advantage to the full, the Finnish pilots downed seven Soviet fighters in just ten minutes, with 1Lt Iikka Törrönen claiming two in BW-385. Forty-eight hours later the 3rd Flight destroyed three SBs and a MiG-3 over the River Svir, between Lake Onega and Lake Ladoga.

On 23 September it was the 3rd Flight's turn to add to its growing score when Capt Karhunen led eight Brewsters in an attack on three I-16s of 155.IAP that were caught strafing troops south of Petrozavodsk, close to

1Lt Pekka Kokko, 3/LLv 24's deputy leader, flares out above the runway at Immola in BW-379 'Orange 9' just prior to landing. Although not visible in this photograph, BW-379 carried a rare personal marking in the form of Kokko's christian name just aft of its cowling on the starboard side only. A three-kill veteran of the Winter War, Kokko had claimed an additional 6.5 victories in BW-379 by the time this shot was taken in mid September 1941. Two months later, on 24 November, Pekka Kokko was posted to the *Koelentue* (Test Flight) to serve as a test pilot, having increased his score to 13.5 victories claimed during the course of 150 combat sorties (*SA-kuva*)

Lake Onega. Enjoying odds firmly stacked in their favour for a change, the Finnish pilots made short work of all three Polikarpov fighters. Then, following a pre-briefed plan, Karhunen ordered his flight (by radio) to return home while he circled for a further 30 minutes at low level over the wilderness in complete radio silence. He then returned to the spot where the two 'Ratas' had been destroyed and surprised a further six I-16s of 155.IAP that were strafing the infantry once again. Only one escaped.

Three days later Capt Karhunen repeated the same tactic, and this time his flight destroyed all six 'Chaikas' that were initially encountered. Circling at tree-top height for a short while, the Finnish pilots returned to find three I-16s, three I-15bis and two I-153s of 65.ShAP harassing the troops. A further three fighters were downed and the rest fled. Capt Karhunen claimed two in BW-366 and WO Juutilainen three in BW-364. This extract is taken from the latter pilot's combat report;

'During the first combat, which took place just north of a small landing ground, I observed one I-153 coming towards me just above the ground with its gear extended. After spotting me, the pilot started to turn away, but I fired a short burst at his fighter whilst it was banking and it hit a tree and crashed into the woods, which caught fire. 1Lt Kokko saw the whole incident.

'During the second encounter I observed two I-15bis which were trying to escape. One crashed into Lake Onega after being shot down by MSgt Nissinen, whilst the second fighter dived into cloud, and I followed it. Taking its pilot by surprise, I fired into the fuselage of his machine from behind, and he went into a dive. I stayed with the Polikarpov until I saw it crash vertically into the forest. MSgt Nissinen also observed the crash.

'I then wrestled for an advantage with three I-16s, before finally getting on the tail of one of them and shooting it down into woodland next to a road packed solid with troops. Capt Karhunen observed the downed fighter after it had burst into flames on the ground.'

Following the capture of Olonets and Petrozavodsk by 1 October, the Karelian Army advanced slowly northward along the west coast of Lake Onega. Once at the northern tip of the lake, troops occupied the towns of Karhumäki on 5 December and Poventsa the following day, thus ending the Finnish advance. From these defensive positions, a two-and-a-half-year stationary war now began.

Seen at Immola on the same day as Pekka Kokko's BW-379, 4/LLv 24's BW-394 'Black 6' was photographed just prior to taking off. It was regularly flown during this period by 2Lt Erik Teromaa, who would later become both a flight leader and 19-kill ace (although he claimed no victories in this particular Brewster). BW-379 was written off after crash-landing with battle damage on 6 June 1942. Its pilot, 1Lt Uolevi Alvesalo, escaped from the wrecked fighter without having suffered any serious injuries (*SA-kuva*)

On 7 October, whilst supporting this new offensive, six Brewsters headed by Capt Sovelius (in BW-378) had encountered 15 I-153 and I-16 fighters over Kontupohja and shot down five of them. Eight days later Capt Eino Luukkanen was leading a combat air patrol of six Brewsters (in BW-375) over the upper course of the River Svir when;

'Two bombs exploded in the middle of the river, sending two columns of water high up into the air. Tracer rounds from an anti-aircraft battery soon allowed me to pick out the three attacking bombers, which were surrounded by flak burst. Upon closing to firing range, I could not believe my eyes when I saw that all three aircraft were fitted with skis! Firstly, I had never previously seen a twin-engined aeroplane with skis, and secondly, there was no snow to be seen for as far as they eye could see.

'2Lt Kai Metsola (in BW-390) shot the first bomber down in flames into a forest south of Osta, and the second aircraft suffered an identical fate after it was attacked by WO Viktor Pyötsiä (flying BW-376).

'The third machine tried to escape to the east, but I soon caught up with it. Closing to within 100 metres before opening fire, and I saw hits on the fuselage. Suddenly, a vertical, twisting and burning ribbon appeared between me and my target, and I was forced to break off my attack. I had not seen anything like this before, and despite subsequently trying to find out what this burning substance was, it still remains a mystery to me.

'Pulling in behind the bomber once again, I noticed that one of its engines was now smoking, so I aimed for the still operable powerplant

Having had its engine cover removed, BW-375 'White 5' is prepared for its next mission in 1/LLv 24's open dispersal area amongst the trees at Nurmoila, on Olonets Isthmus, in October 1941. The fighter was assigned to 1st Flight leader Capt 'Eikka' Luukkanen at the time, and he used it to claim 4.5 kills between 8 July and 7 November 1941. His personal marking consisted of a white stripe on the fighter's blue spinner, as seen here – blue was also the 1st Flight colour, and was repeated on the Brewster's rudder

1/LLv 24's BW-382 'White 9' sits next to the servicing hangar at Nurmoila in October 1941. Ex-2nd Flight Winter War veteran WO 'Veka' Rimminen was issued with BW-382 upon his assignment to the 1st Flight in early August 1941. Having already claimed two and one shared destroyed with 2/LLv 24 (one and one shared were scored on D.XXIs in 1940), Rimminen was credited with one and two shared destroyed in BW-382 in August-September. He continued to fly 'White 9' until he was posted to the Air Fighting School to serve as an instructor on 15 September 1942, by which time his score stood at 5.333 kills

instead, and it also started to emit smoke. Now just 50 metres behind the bomber, and in a shallow dive, I spotted two darkly-clothed individuals jump out of the stricken aircraft. Upon pulling up and away from the now pilotless bomber, I saw two parachutes drifting down. The aeroplane continued its dive until it crashed into woodland east of Sulandozero.'

These SBs belonged to 72.SBAP, and had obviously flown down from bases in the north, where snow had already fallen – worsening winter weather would restrict flying on both sides for the next two months.

On 17 December the last big combat of 1941 took place when four Brewsters, led by Capt Karhunen, encountered a mixed gaggle of nine Hurricanes and 'Chaikas' during the course of a patrol west of the White Sea. The engagement lasted 20 minutes, and five Soviet fighters were shot down – Karhunen claimed a single example of each type in BW-366.

By 23 December LLv 24's headquarters, 3rd and 4th Flights had moved to the seasonal airstrip near the small port of Kontupohja, situated on the west coast of central Lake Onega – it was deemed to be 'seasonal' because the town's frozen bay within the port served as the runway!

During 1941, the unit had claimed 135 aerial victories without suffering a single loss to enemy fighters. Indeed, its only operational fatality had been caused on 3 December when 1Lt Henrik Elfving was shot down by Soviet anti-aircraft fire at Novinka whilst flying BW-385.

LLV 24's outstanding record again made it the best fighter unit within the Finnish Air Force. By comparison, LLv 26 was credited with 52 kills without loss flying Fiat G.50s, LLv 28 had scored 70 victories with its Morane-Saulnier MS.406s, but suffered five casualties in the process, and LLv 32 destroyed 52 aircraft with its Curtiss Hawks, and also lost five pilots to enemy fighters.

Looking very much like a scaled-down Morane-Saulnier MS.406, but with a radial engine and two seats, the State Aircraft Factory designed and constructed 51 VL Pyry (Whirlwind) advanced trainers for the Finnish Air Force in 1940-41. Most frontline squadrons were assigned one or two examples for evaluation purposes, and PY-33 served with LLv 24 between 29 April and 5 July 1941. Issued to the Air Fighting School following its brief time with the air force's premier fighter unit, PY-33 was photographed at Kauhava on 22 October 1941 in an identical scheme to the one that it wore whilst attached to LLv 24 (*Finnish Air Force*)

STATIONARY WAR

In August 1941 Lend-Lease shipments of ex-RAF Hurricanes from the UK started to arrive in significant numbers at the Soviet Artic ports of Murmansk and Arkhangelsk. Once reassembled, these fighters were primarily used against the Germans by the Northern Fleet air forces in the Murmansk and Kandalaksha areas. However, as more aircraft arrived in the USSR, regular air force units along the Finnish border also started to re-equip with Hurricane IIAs and IIBs.

With British-built fighters starting to reach units south of Murmansk, LLv 14 and its obsolescent Fokker D.XXIs prepared for action from the Finnish Air Force's northernmost base at Tiiksjärvi, 200 km west of Belomorsk on the White Sea. Senior officers within the air force were fully aware that the unit had little hope of repelling ever-increasing numbers of more modern Soviet fighters on its own, so on 8 January 1942 2/LLv 24 was temporarily deployed to Tiiksjärvi with eight Brewsters. Within two weeks, it had been strengthened by the arrival of another four aircraft.

Exactly a week prior to this move northwards, LLv 24's disposition was as follows;

***Lentolaivue* 24 on 1 January 1942**

Commander	Lt Col Gustaf Magnusson, with HQ at Kontupohja
1st Flight	Capt Eino Luukkanen at Nurmoila with seven Brewsters
2nd Flight	Capt Leo Ahola at Helsinki-Malmi with six Brewsters
3rd Flight	Capt Jorma Karhunen at Kontupohja with eleven Brewsters
4th Flight	Capt Per-Erik Sovelius at Kontupohja with eight Brewsters

On 24 January 2/LLv 24 engaged the enemy for the first time in the Rukajärvi area, when five Brewsters ran into ten I-15bis and I-153 fighters from 65.ShAP and claimed four of them shot down. A fifth Soviet

WO 'Illu' Juutilainen of 3/LLv 24 taxies BW-364 'Orange 4' past the dockside facilities at Kontupohja harbour in early February 1942. Most Brewsters operating from this site were not fitted with skis, as the ice runway was usually ploughed and rolled flat on a daily basis. The aircraft's distinctive Continuation War victory silhouettes on its fin were introduced as the standard air force kill marking in September 1941

A Brewster is seen undergoing routine maintenance within Kontupohja's snug-fitting servicing hangar on 15 February 1942. This heated facility was hastily erected in order to provide shelter for mechanics struggling to work in temperatures as low as -30°C. The base at Kontupohja was comprised of modest lakeside harbour buildings and an ice runway on the frozen bay (*SA-kuva*)

fighter was also destroyed when SrLt V A Knizhnik deliberately rammed his I-153 into Sgt Paavo Koskela's BW-372, the 'Chaika' forced-landing and the Brewster returning to the base with a broken starboard wing tip – both pilots claimed an aerial victory! However, the unit did lose one of its precious Model 239s later in the day to a more conventional attack when Sgt Eino Myllymäki, in BW-358, was reported missing from a bomber escort mission to Belomorsk. Recent research has confirmed that he was shot down by a Hurricane from 152.IAP.

As the conflict broadened along the Finnish border, other flights within LLv 24 started to extend their patrols from Kontupohja to as close to the White Sea as their fuel tanks would allow them. Whilst flying just such a mission, on 26 February, the 3rd Flight encountered MiGs over the Isthmus of Maaselkä, between the White Sea and Lake Onega. Capt 'Joppe' Karhunen, flying BW-366, shot two of them down. His combat report read;

BW-388 'Orange 5' of 3/LLv 24 sits parked alongside a more conventional means of conveyance at Kontupohja harbour on 15 February 1942. This particular aircraft was assigned to the flight's deputy leader, 1Lt Osmo Kauppinen, who scored 5.5 kills (two in BW-388) in 67 sorties. The dock installations can been seen behind the Brewster (*SA-kuva*)

'Whilst on a search mission we engaged 15-17 MiG-1s and MiG-3s at a height of 2500 metres over Liistepohja. The first phase of the battle took place close to the frontline, whilst the second phase was fought to the west of Juka and Nopsa stations, on the Murmansk railway line.

'I hit one MiG-1 from behind and it made two fast rolls. I then hit it again, and the pilot pulled the fighter's nose up a little before entering a spiralling dive down to 800 metres, where I broke off my pursuit and continued the fight with other MiGs. After a brief turning battle with a MiG-3, I succeeded in hitting the fighter in the engine, and its pilot made a forced landing with a windmilling propeller some four kilometres north-east of Nopsa station.

'I fired at five aircraft in all. The MiGs were split into 4-6 aeroplane divisions, which were flying in pairs, staggered about 500 metres apart. They were well above us when we first spotted them, but they did not exploit their altitude advantage.'

LLv 24 did not escape unscathed, however, for Sgt Tauno Heinonen was forced to bail out of BW-359 at 1020 hrs when his fighter was shot up by a MiG-3 over Listepohja. Fortunately for him, he landed about one kilometre inside the Finnish forward lines. There is a possibility that these MiGs were in fact Hurricanes from 152.IAP.

Capt Karhunen repeated this mission on 9 March, when he led eight Brewsters on a 'free hunt' in the direction of Segesha. On the way to the target area, a bomber and a fighter were met and quickly shot down. Nearing Uikujärvi, the Finns were bounced by six Hurricanes of 152.IAP, and in the ensuing combat three Soviet fighters were shot down for the loss of BW-362, flown by Sgt Paavo Mellin.

A 5.5-kill ace, who had scored his all-important fifth kill as recently as 26 February (in BW-362), Mellin bailed out of the striken fighter and landed in thick snow. After wading all day through drifts up to his waist, Mellin lost consciousness from exhaustion west of the Murmansk railway line. Fortunately for the 22-year-old pilot, a Russian scouting patrol returning from Finnish territory found Mellin and saved him from freezing to death. He eventually returned home in December 1944 as part of the exchange of the PoWs that took place between Finland and the USSR.

SUURSAARI OPERATION

The island of Suursaari is located south of Kotka in the middle of the Gulf of Finland. It was held by the Russians until early December 1941, when

Covered with tarpaulins, two of 4/LLv 24's Brewsters sit quietly within the unit's shoreline dispersal area at Kontupohja on 15 February 1942. In the foreground is BW-380 'Black 1', which was assigned to 1Lt Iikka Törrönen, whilst BW-378 'Black 5' belonged to flight leader, Capt 'Pelle' Sovelius. Twenty-four hours after this photograph was taken, the latter pilot was posted to air force HQ, and Törrönen took over the flight (*SA-kuva*)

they evacuated their base in the face of Finnish advances. However, a subsequent realisation of its strategic importance saw the Red Army re-occupy the island on 2 January 1942. This action in turn prompted the Finnish Army to attempt to seize the island back whilst its troops could still safely advance over the ice.

Accordingly, on 27 March, a 3500-man occupation force started the advance on Suursaari, supported by some 57 aircraft. LLv 6 provided five SB bombers and six I-153 fighters (all of which had been captured in either the Winter or Continuation Wars), LLv 24 committed six Brewsters, LLv 30 16 D.XXIs, LLv 32 13 Hawks and LLv 42 11 Blenheims.

On the day of the invasion the Finns shot down four defending fighters, and 24 hours later two more large-scale engagements took place. At 0800 hrs 1Lt Osmo Kauppinen led six Brewsters into battle against ten I-153s of 71.IAP, and his flight duly claimed half of them shot down – WO 'Illu' Juutilainen (in BW-364) and Sgt 'Jussi' Huotari (in BW-353) were each credited with two apiece.

The island was successfully recaptured by Finnish troops within hours of the invasion commencing, and whilst an impromptu victory parade took place on the ground in the late afternoon, overhead 12 Hawk pilots from LLv 32 clashed with 29 Soviet aircraft from 11. and 71.IAP. During a 20-minute melee, the Finns claimed to have shot down ten I-153s and five I-16s without loss – Soviet loss records admit the destruction of one I-15bis, one I-16 and six I-153s.

LLv 24's 'Illu' Juutilainen had now achieved 20 aerial victories during the Continuation War (to add to his two and one shared destroyed from the Winter War), giving him a score comparable with LLv 26's 'ace of aces', WO Oiva Tuominen. He was rewarded for this success through the receipt of the Mannerheim Cross (the 56th awarded to a member of the Finnish armed forces). The first pilot within his unit to be honoured with

MANNERHEIM CROSS

Following the end of the Winter War, the statute governing the creation of the Mannerheim Cross was issued on 16 December 1940. The recipient became a Knight of the Mannerheim Cross, and it was awarded in two classes for extraordinary bravery – highly notable achievements in battle, or outstanding leadership in combat. Its awarding was not dependent on the rank of the nominee.

The first cross (2nd class) was issued on 22 July 1941 to Col Ernst Lagus for heading armoured units in battle. The first air force Mannerheim Cross Knight (award number six) was LLv 26's WO Oiva Tuominen, who received his award on 18 August 1941 for having scored 20 kills, eight of them in the Winter War. All future recipients from the air force had only their present achievements taken into account.

The Mannerheim Cross can be compared with the British Victoria Cross or the American Medal of Honor, being the highest military award in Finland. Some 191 were issued, 19 of which were presented to air force pilots. Only four men received it twice – fighter pilots Capt Hans Wind and WO Ilmari Juutilainen on 28 June 1944, and Maj Gen Aaro Pajari and Col Martti Aho on 16 October 1944.

The Mannerheim Cross 1st class was issued only twice, to Marshal of Finland Carl Gustaf Emil Mannerheim on 17 October 1941 (award number 18), and to Chief of Staff Gen Axel Heinrichs on 31 December 1944. Heinrichs had earlier received the 2nd class award (number 48) on 5 February 1942.

The award also brought with it the sum of 50,000 Finn marks, an amount then equalling the annual salary of a regular first lieutenant in the armed forces.

BW-362 'Orange 2' of 3/LLv 24 idles at Kontupohja prior to taking off. Within days of this photograph being taken, 5.5-kill ace Sgt Paavo Mellin was shot down in this aeroplane on 9 March 1942. He bailed out and was captured, remaining in captivity until Christmas 1944. BW-362's temporary winter whitewash consisted of glue and chalk mixed together, which was then brushed on in whatever pattern took the groundcrewman's fancy! (*R Lampelto*)

Ilmari Juutilainen opens BW-364's throttle as he prepares to taxi out across the frozen runway at Kontupohja in early March 1942. The photographer is actually standing on the wharf, looking down into the harbour. The groundcrewman responsible for whitewashing this particular Brewster has tried to cover the black areas of the standard camouflage pattern, as well as most of the aircraft's serial and rudder number (*R Lampelto*)

Finland's highest military honour, Juutilainen's decoration was officially presented on 26 April 1942.

Twelve days prior to receiving his award, Juutilainen and his fellow pilots from the 3rd and 4th Flights – as well as LLv 24's HQ staff – had vacated the thawing ice of Lake Onega and flown to the permanent base at Hirvas, located in the middle of Karelian wilderness.

FIGHTING LEND-LEASE

During February 1942 the Tiiksjärvi-based Brewsters of 2/LLv 24 had twice engaged Hurricanes, claiming three destroyed on both occasions, but March proved to be far quieter. However, this period was very much the 'lull before the storm', for senior Red Army officers had decided to

BW-374 'White 4' was assigned to 1/LLv 24 at Nurmoila in March 1942. Flown by long-time flight member SSgt Aimo Vahvelainen, the fighter has had its wheels replaced with retractable skis so as to allow it to cope better with snow-covered airfields during the winter months. This modification was only carried out on a few Brewsters, with the vast majority retaining their wheels and tyres all year round (*E Luukkanen*)

eliminate the Finnish Air Force's sole permanent airfield in the north-east once and for all.

The opening phase in this campaign came on 29 March, when seven Hurricanes strafed Tiiksjärvi. In response, 1Lt Lauri Pekuri (in BW-372) led eight Brewsters on a reconnaissance mission to Segesha 24 hours later. Some 12 Hurricanes from 152.IAP were duly bounced during the sweep, and six were shot down.

Eight days later the Red Air Force carried out its long-planned air raid on Tiiksjärvi, the 26-strong (14 bombers and 12 fighters) formation being detected by Finnish radio intelligence well short of the objective. Their exact position was relayed to 1Lt 'Lasse' Pekuri, who was leading a formation of eight Brewsters that were conducting a routine patrol at the time. The Finnish fighters attacked just minutes prior to the Russians reaching Tiiksjärvi, and two 'DB-3' bombers (actually SBs of 80.BAP) and twelve Hurricanes (from 609 and 767.IAP) were shot down without loss between 1525 and 1550 hrs.

1Lt Pekuri (in BW-372) and 2Lt Nissinen (in BW-384) both claimed three apiece, while MSgt 'Lekkeri' Kinnunen (flying BW-352) destroyed two, taking his score to 16. The latter pilot's combat report recounted;

SSgt Aimo Vahvelainen was at the controls of BW-371 'White 1' when this accident occurred during his take-off run from Nurmoila on 12 April 1942. The 1/LLv 24 machine got its undercarriage bogged down in the melting slush as the pilot accelerated down the runway, yawed uncontrollably to the right and then hit a snow bank and flipped over. Aimo Vahvelainen escaped without injury, and BW-371 was trucked off to the State Aircraft Factory to be repaired. It was damaged once again during a post-rectification check flight, and did not return to LeLv 24 until 13 February 1943

'I flew in the top swarm, leading the second pair. We engaged an enemy formation comprised of seven bombers and eighteen fighters. I attacked a formation of six Hurricanes flying behind the SBs, and that is when the real dogfighting began. I fired at several fighters during this brief engagement, and one of them dropped away trailing a thick plume of smoke, before half-rolling into a vertical dive and crashing straight into a forest.

'I then chased after a second Hurricane, closing to within 100 metres of its tail before opening fire. Reducing the distance of separation to just 50 metres, I fired again and the fighter began to smoke, catching fire on the underside of the fuselage. Entering a shallow dive, the Hurricane eventually crashed into a small field and was totally destroyed.'

Official Soviet losses are known to have amounted to one SB from 80.BAP, two Hurricanes from 609.IAP and a further four from 767.IAP. In return, the communists claimed four aircraft destroyed on the ground and seven Brewsters in the air. In reality, only one Model 239 (BW-394) was damaged by return fire from an SB, its pilot, 1Lt Lasse Kilpinen, suffering a serious calf wound – he still shot the Tupolev down, however.

Following the poor results of this raid, the Red Air Force was not seen again until 8 June, when 1Lt Pekuri's six Brewsters engaged thirteen Hurricanes from the Kesä-based 152.IAP. Five Soviet fighters were destroyed, but on this occasion a Model 239 was also lost.

1Lt Uolevi Alvesalo's BW-394 had been shot up during the melee, and he succeeded in limping back as far as Rukajärvi, before carrying out a textbook forced landing. He escaped without injury. This was the second time Uolevi Alvesalo had walked away from a wrecked Brewster, for on 29 January, whilst testing the guns of BW-389, he had struck the ice in a shallow dive at a speed of 450 km/h. Astoundingly, he emerged from the shattered remains of his fighter with little more than concussion!

2Lt Lauri Nissinen had also claimed one kill – in BW-384 – during the 8 June action, taking his tally to 20, and earning him LeLv 24's (the abbreviation of *Lentolaivue* changed from LLv to LeLv in May 1942) second Mannerheim Cross on 5 July 1942. Recipient number 69 of Finland's highest military medal, reservist Nissinen was notified of the award soon after enrolling in the air force's cadet school, where he would train to be a regular officer.

Another of the handful of ski-equipped Model 239s, BW-358 'White 1' has its engine run up in preparation for take-off at 2/LLv 24's Tiiksjärvi base in early 1942. This aircraft was shot down in combat near Sorokka (Belomorsk), on the the White Sea coast, by Hurricanes of 152.IAP on 24 January 1942. The Brewster's pilot, Sgt Eino Myllymäki, was posted missing in action

SSgt 'Hemmi' Lampi inspects work carried out by his mechanics on BW-354 'White 6' at Tiiksjärvi in early April 1942. The 2/LLv 24 pilot scored the Finnish Air Force's first victories of the Continuation War in this very fighter when he downed two and one shared SB-2bis bombers on 25 June 1941. The bottom two kill markings displayed on the fighter's fin denote two Hurricanes that Lampi had destroyed in BW-354 on 30 March 1942. Later promoted to second lieutnant, Heimo Lampi finished the war with 13.5 kills. Unlike its pilot, BW-354 did not survive the conflict, being shot down by an La-5 of 4.GIAP, KBF over Oranienbaum on 21 April 1943. SSgt Tauno Heinonen was killed in the action (*K Temmes*)

On 25 June the last big combat of the Tiiksjärvi campaign took place north-east of Lake Seesjärvi, when swarms from both 2 and 3/LeLv 24 engaged in a 15-minute fight with Hurricanes from 609.IAP. The latter flight claimed four Hurricanes without losses, whilst the 2nd Flight got three, but also lost two Brewsters. 1Lt Lauri Pekuri ditched his burning BW-372 into a small lake in the wilderness behind enemy lines and walked 20 km back to Finnish territory. Sgt Kalevi Anttila (in BW-381) also crashed-landed in the same area, as fellow 2/LeLv 24 pilot WO Yrjö Turkka explains;

'It is a hot and sunny day, and Anttila is in the scramble swarm. The pilots have discarded their flying suits and are sunbathing next to their fighters, wearing just swimming trunks, when the alarm siren sounds.

'"ALARM, ALARM, fighters spotted approaching Rukajärvi".

'Air force regulations state that a pilot has to wear the appropriate flying gear on all missions, but Anttila thinks that he can get away without wearing his suit on this occasion, for he assumes that the engagement will take place above Tiiksjärvi. He therefore grabs just his pistol in its holster, slings it around his neck, and then climbs into the hot cockpit of his Brewster fighter.

'Once realising that they have been detected, the enemy fighters turn back east, so formation leader, 1Lt Pekuri, orders the flight to chase after them in typical reconnaissance detachment fashion. The chase continues as far as the main Soviet air base at Segesha, when, all of a sudden, the Brewsters are bounced by Hurricanes. Anttila's aeroplane is immediately hit in the engine, causing it to stop.

'He is now forced to weigh up his options. He cannot land alongside the Murmansk railway line, which runs immediately below him, for his aeroplane will almost certainly catch fire, thus giving away his position. He cannot bail out, for descending down into thick forest bare-footed, and clad in only swimming trunks, will result in him suffering cuts and abrasions. Then, he observes a small swamp, and immediately belly-lands

BW-384 'Orange 3' is seen wrapped up against the elements at Tiiksjärvi in mid February 1942. This aircraft was flown by one of LLv 24's top scorers in MSgt Lauri Nissinen. Although transferred to the 2nd Flight with its pilot on 28 January 1942, the fighter continued to wear its old 3rd Flight colours until well into the spring. BW-384 was also one of the few aircraft at Tiiksjärvi to lack the squadron's lynx emblem

'Lapra' Nissinen sits on the tailplane of BW-384 in late February 1942, the fighter's fin displaying silhouettes for all 15.5 of his Continuation War victories scored up to that point in the conflict – the ace's four Winter War kills are not marked. On 8 June 1942 Nissinen destroyed his 20th aircraft (12.5 of which were downed in BW-384), and he was duly awarded the Mannerheim Cross on 5 July 1942

the aeroplane along the water's edge, deliberately severing its wings by hitting small pine trees whilst skidding to a halt.

'Anttila suffers no injuries in the landing, and prior to leaving the fighter, he smashes its instruments and radio with the handle of his pistol. Tucking his parachute and maps under his arm, he starts running in the direction of friendly territory. He soon crosses a small creek, and proceeds to circle around it several times in the hope of shaking off the scent of any Russian search parties looking for him with dogs. Pressing on upstream, Anttila fords through the water for an hour before stopping to rest, and take stock of his predicament.

'He knows that he is north of Lake Lintujärvi, at least 60 km from the Finnish frontline, and that the ground between him and safety is covered by swamps and thick forest. Unsure as to how long it would take him to skirt around all these obstacles, Anttila is thankful that he at least bothered to pick up his pistol before taking off, for its holster contains a compass. Looking down at his bleeding bare feet, and the thousands of mosquitoes swarming around him, he briefly considers returning to the nearby Murmansk railway line to surrender, then immediately banishes the thought from his mind.

'Instead, Anttila climbs the nearest hill to see if anybody is following him, and then heads off in the direction of the Finnish frontline at Ontajärvi. He cannot sleep during the night due to the swarms of mosquitoes and the plummeting temperature, so he continues walking in order to stay warm. In the morning he rests, sitting on a fallen tree as the sun begins to warm him up. He cannot sleep, however, thanks to the swarming mosquitoes, so he forces himself on until he arrives at a swamp. Realising that the moss which surrounds it appears to be dry, Anttila digs himself a hole, lays down in it and pulls more dry moss over him. He falls asleep with the pistol in his hand.

'The next night he continues his trek across swamps, often sinking down into water that rises up to his waist, and expending a lot of energy getting back onto his feet. Exhausted by this effort come the following morning, and having not eaten anything, Anttila decides not to dig a hole into the moss again, fearing that he would not wake up from the resulting sleep.

'Coming to the edge of yet another swamp, he suddenly sees something in the middle distance. With his vision impaired through fatigue, he has trouble making the object out – is it a wolf, or a pursuer? Is it all going to end here? No, it's not a man, nor a wolf. Anttila draws his pistol and starts to approach. It's a young moose staring at him. He gets closer and the animal just stares at him. He tries to shoot it from a distance of ten metres, but his hand trembles too much, and he forces himself closer. At two metres he fires the pistol and hits the baby moose. Blood begins to bubble up from a gaping throat wound.

On 24 January 1942 2/LLv 24's Sgt Paavo Koskela, in BW-372, snapped off the interplane struts of a ramming 65.ShAP (assault aviation regiment) I-153, causing its pilot, SrLt V A Knizhnik, to force-land. Koskela, meanwhile, coaxed his badly-handling BW-372 back to Tiiksjärvi, where he landed and then collapsed from total exhaustion. Both pilots claimed an aerial victory, although this photograph clearly proves that only the Finn should have been credited with a kill. Once repaired, BW-372 remained in frontline service until damaged in combat by a Hurricane from 609.IAP near Seesjärvi on 25 June 1942. Its pilot, 1Lt Lauri Pekuri, was forced to ditch the burning fighter into a small lake behind enemy lines, and although the Brewster was lost, the 18.5-kill ace managed to escape back to Finnish territory. Pekuri had claimed seven victories in this machine between 30 March and 25 June, including two Hurricanes just minutes prior to BW-372's demise

'Realising that his salvation is at hand, Anttila forces himself to suck the blood from the moose. His stomach is not prepared for anything like this, and it takes all his strength to stop himself from vomiting. Soon, his hunger, and will to survive, take over, and he continues to suck. His strength returns thanks to the infusion of blood, and he continues his escape westward. That evening Anttila makes it to a road running between Paatene and Rukajärvi, where he eventually waves down an approaching car, and almost scares its occupants to death – he has the appearance of a naked savage, covered in blood and dirt.

'The telephone rings at our base and is answered. Unbelievably, "Serg" Anttila, the dead man, has come back.'

In six months of combat, the Brewster pilots of the reinforced 2nd Flight had claimed 45 Hurricanes shot down. Such a tally is perfectly feasible, bearing in mind that the average strength of a Soviet fighter regiment (IAP) was 15-20 serviceable aircraft during this period. Their job done, the pilots of 2/LeLv 24 would see little action from Tiiksjärvi until November 1942.

REORGANISATION

These final engagements in June signalled the start of six months of relative stability along the Soviet frontlines, which allowed senior staff officers within the Finnish Air Force to implement plans for a radical reorganisation of its fighter squadrons. Changes were made in order to

Ex-RAF (formerly used by No 401 'Canadian' Sqn) Hurricane IIB Z3577 was shot down on 6 April 1942 near Tiiksjärvi when eight Brewsters from 2/LLv 24 attacked 14 bombers and 18 fighters. The Finns claimed 12 Hurricanes, and both 609. and 767.IAP admitted losing aircraft during the battle. This particular fighter displays damage caused by both flak and Brewster machine gun fire (*SA-kuva*)

Well-weathered 4/LLv 24 machine BW-386 'Black 3' is seen at Immola in April 1942, just prior to undergoing a much-needed overhaul. The fighter was assigned to MSgt Sakari Ikonen for much of the Continuation War, the ace claiming his sixth, and last, victory with BW-386 on 22 November 1942 (he had also scored two other kills with it in 1941). Shortly after 'making ace', 204-sortie veteran Ikonen – who had also fought in the Winter War – was posted to the Air Fighting School as an instructor. The 4th Flight's osprey emblem is clearly visible on BW-386's fin, this marking rarely being seen on Model 239s. Indeed, it disappeared altogether following the disbanding of 4/LeLv 24 on 15 February 1943 (*E Laiho*)

improve the country's area defence system, although air regiment commanders 'in the field' considered that the new strategy would see their units losing an element of combat flexibility – the main tactical advantage employed so effectively by Finnish fighter pilots to date.

Nevertheless, on 3 May 1942 the front was divided into three sectors, with a single regiment then being given the responsibility of defending the airspace within that particular sector. The Lake Onega sector was covered by LeR 2, comprising LeLv 16, 24 and 28, Olonets belonged to LeR 1, with LeLv 12 and 32, and the Karelian Isthmus remained with LeR 3, controlling LeLv 26 and 30. The northern flank was covered by LeLv 14, whilst in the south LeLv 6 handled the Gulf of Finland. Bomber regiment LeR 4 was moved to which sector needed it most at the time.

As a result of this reorganisation, LeLv 24's Nurmoila-based 1st Flight joined the bulk of the unit at Hirvas on 31 May, which in turn moved to Suulajärvi, on the Karelian Isthmus, ten days later. Here, it was placed directly under the command of LeR 3's CO, Lt Col Einari Nuotio.

WO Ilmari Juutilainen poses with the Mannerheim Cross pinned to his chest (on his left breast pocket) in front of BW-364 at Hirvas on 26 April 1942 – the very day the medal was presented to him. He was the first LLv 24 pilot to receive Finland's highest military decoration, and it was awarded to him for scoring 20 Continuation War victories – 16 of them in BW-364 (*R Lampelto*)

The new system received further adjustment on 18 July 1942 when the rest of LeLv 24 was transferred to LeR 3. More changes occurred on 16 November following LeR 5's formation to oversee the maritime patrol operations of LeLv 6 and 30. And finally, on 23 January 1943, a new fighter unit in the form of LeLv 34 was established for LeR 3.

GULF OF FINLAND

When the ice melted in the Gulf of Finland in May 1942, the Red Banner Baltic Fleet started sending out submarines into the coastal shipping lanes from its huge naval (and air) base at Kronstadt, on the outskirts of Leningrad. These vessels had retreated into port the previous autumn, and with the spring thaw, their sole purpose was to harass German and Finnish commercial shipping in the Baltic Sea. The early summer also saw the Red Banner Baltic Fleet air force increased in size so as to be able to cover naval movements in the area – especially the submarine arrivals and departures to and from Kronstadt, in the eastern Gulf of Finland.

To counter this growing aerial activity, LeLv 24 was seconded to LeR 3 on 18 July 1942. Unit strength then stood at 27 serviceable Brewsters, seven of which were still defending the northern sector with the 2nd Flight from Tiiksjärvi. LeR 3 welcomed the squadron headquarters staff and the 3rd and 4th Flights at Römpötti air base, on the Karelian Isthmus, on 1 August – the 1st Flight flew in one week later. Their role was to prevent enemy aircraft from flying over the western areas of the gulf, although the Russians chose to fly under the protection of anti-aircraft batteries in the vicinity of Oranienbaum (35 km west of Leningrad). With Soviet aircraft now flying outside of the Brewsters' effective range, no large-scale encounters took place.

'Illu' Juutilainen runs up BW-364's Wright Cyclone engine within 3/LeLv 24's blast pen at Hirvas in May 1942. The last two kills of the twenty victory bars on the fighter's fin denote a pair of 'Chaikas' (from 71.IAP, KBF) shot down during the invasion of Suursaari, in the Gulf of Finland, on 28 March 1942. Juutilainen would go on to claim a further 12 kills with this fighter (plus single victories in BW-368 and BW-351) by the end of November

2/LeLv 24's BW-387 hangs suspended from the ceiling of the servicing hangar at Onttola in June 1942. It has been hoisted up to allow a series of recognition photographs to be taken for official air force and army recognition manuals. The repetition of the fighter's serial on the wing undersurfaces was a distinguishing feature of the Brewsters reassembled in Sweden during the Winter War (*Finnish Air Force*)

Nonetheless, some engagements did occur, with the first of these being fought on 6 August near Seiskari when the 1/LeLv 24 downed two I-16s. Six days later the 4th Flight got an Il-2 and an I-16 over Tolli lighthouse.

Much of LeLv 24's intelligence on their counterparts' activities at this time was derived from a forward observation post at Ino, where, on a clear day, 'spotters' could actually see Soviet aircraft taking off and landing at Kronstadt and Oranienbaum! The post was usually manned by a pilot (officer) from the unit, and he could guide the scrambled Brewsters to their target through radio transmissions. Armed with this advanced warning of air activity, LeLv 24 started to employ new tactics whereby the Brewsters would be sent aloft to await the return of Soviet aircraft, which were low on fuel and ammunition after having completed their mission against German forces in Estonia.

These sorties soon proved productive, for in two separate missions flown on 14 August, Finnish pilots claimed nine Hurricanes shot down. Forty-eight hours later, the 3rd Flight engaged a large formation of enemy aircraft. Flight leader, Capt 'Joppe' Karhunen (in BW-388), led the attack, which took place at 1745 hrs;

'I was leading a six-aeroplane Brewster flight on an interception mission when, south of Seiskari, I spotted an enemy formation of eight SBs, three

BW-387 'Black 8' was assigned to 2/LeLv 24 at Tiiksjärvi in May 1942. SSgt Aarne Korhonen used this aircraft to down all four of his victories during this period, the pilot just missing out on becoming an ace prior to being posted to a maritime patrol squadron. The fighter is marked with the 4th Flight's distinctive white rudder and black number, although the 2nd Flight's elk emblem can just be made out on the fin (*Finnish Air Force*)

Well-used Brewster BW-378 'Black 5' of 4/LeLv 24 was photographed during a visit to Vesivehmaa in May 1942. When its regular pilot, Capt Sovelius (who had scored seven victories with it), was transferred to air force HQ on 16 February 1942, this machine was assigned to 1Lt Hans Wind (who only managed to claim two shared victories with BW-378). In August the fighter was passed on to 2Lt Aarno Raitio, and he was duly shot down and killed in it on the 18th of that month whilst fighting with I-16s of 71.IAP, KBF over Kronstadt (*O Riekki*)

MiGs and sixteen I-16s, which was flying at a height of 200 metres. We attacked the fighter escort, and whilst the I-16s chose to fight, the others fled. In my first dive I shot down the I-16 at the extreme left of the formation, the aircraft diving into the sea. My second I-16 crashed into the water on fire, turning over as it hit. My third I-16 had just evaded the fire from another Brewster when I hit it with several bursts, the fighter then pulling up and being struck again, before falling into the sea wing first. I made 12 attacks.

'The escort flew at the same altitude as the SBs, and although the I-16 pilots fought bravely, they failed to use their numerical strength to pull up on the bombers' flanks and hit us from above. The fighters were armed with cannon, machine guns and rocket projectiles.'

The Brewster pilots claimed 11 I-16s from 4.GIAP, KBF (Guard's Fighter Aviation Regiment of the Red Banner Baltic Fleet), whilst the 'MiGs' encountered were actually Il-2s from 57.AP, KBF.

On 18 August the largest combat of the summer occurred when information was received that 'ten' I-16s had been spotted near Tytärsaari, heading eastwards. 1Lt Hans Wind scrambled with eight Brewsters at 2000 hrs and flew to Seiskari to await the Russians' arrival. However, upon sighting the enemy, it was realised that there were nearer to 60 aircraft in the formation, so Capt Karhunen and 1Lt Lumme both immediately scrambled with their Brewster swarms to offer further assistance.

Undaunted by the odds, the 16 Brewster pilots dived into the Red Banner Fleet. At the controls of his faithful BW-364 (in which he had claimed three kills in the past five days) was Eino Juutilainen;

'We received a frantic message that a large-scale action was taking place east of Seiskari, so my flight scrambled with four aeroplanes and headed for Kreivinlahti, where we soon spotted a vast number of enemy aircraft fighting a handful of Brewsters. I quickly shot an I-16 down within minutes of entering the fray, the fighter crashing into the sea next to patrol boats cruising in the water below us.

'The second I-16 I shot at was hit during a head-on attack, and it banked over and dived straight into the sea south-west of Kronstadt. My third, and final, kill crashed in flames into the water some 500 metres south of victory number two.

'I could not determine the final number of enemy aeroplanes we encountered, there were so many of them.

'Throughout the engagement, gunners aboard the 14-15 Soviet patrol boats below us, in the Tolli lighthouse and in the flak emplacements surrounding Kronstadt and Oranienbaum were firing at us.

Despite being totally outnumbered, LeLv 24 suffered just a solitary casualty – 2Lt Aarno Raitio was shot down and killed (in BW-378) by an I-16 from 71.IAP, KBF. In return, the Finnish pilots claimed two Pe-2s, one Hurricane and thirteen I-16s shot down. In addition to Juutilainen's trio of kills, 1Lt 'Hasse' Wind (in BW-393) and Capt 'Joppe' Karhunen (in BW-388) also sent down three apiece. The Soviets officially reported the loss of at least one Yak-1 and one LaGG-3 from 21.IAP, KBF and two I-16s from 71.IAP, KBF.

Many of LeLv 24's pilots were now entering their second year in combat, having flown the Brewster in action for a full year. With numerous aces boasting scores well into double figures, their inexperienced communist opponents were little more than cannon fodder for LeLv 24 during this period. Within a week the Soviets had lost 39 aircraft.

Forced to make good these losses during the final fortnight of August, no Red Air Force units ventured forth over the frontline until the 31st, when Capt 'Eikka' Luukkanen's flight encountered eight 'Chaikas' near Lavansaari. Four of them were shot down into the sea.

During the month of August LeLv 24 had claimed the destruction of 50 aircraft from the Red Banner Baltic Fleet air force, with half of this tally falling to the guns of Capt Jorma Karhunen's 3rd Flight. His personal score had now reached 20 (plus five from the Winter War), and he was

Brewsters of 1/LeLv 24 rest between sorties at Suulajärvi in June 1942. The fighter closest to the camera is Capt Eino Luukkanen's newly-allocated mount BW-393 'White 7' (which replaced BW-375), and next to it is BW-382 'White 9', flown by fellow ace WO Veikko Rimminen. A Winter War veteran like Luukkanen, Rimminen left LeLv 24 to become an instructor on 15 September 1942, having by then completed 190 sorties and claimed 5.5 kills, three of which were scored in BW-382 (*V Lakio*)

Five Brewsters of 1/LeLv 24 sit out in the open at a sunny Römpötti in August 1942. Flight leader Capt 'Eikka' Luukkanen was assigned BW-393 'White 7' on 1 June 1942 after it had returned to the squadron following major repair work at Tampere – hence its freshly applied camouflage scheme. His previous mount, BW-375 'White 5', is visible immediately behind BW-393. Luukkanen was not a fan of unit or personal emblems on his aircraft, which explains why 'White 7' is the only Brewster in the photograph lacking LeLv 24's distinctive lynx marking (*V Lakio*)

awarded the Mannerheim Cross (number 92) on 8 September 1942. He became the third pilot within his unit to receive Finland's highest honour.

In early September the focus of Russian aerial activity switched to the Isthmus of Maaselkä, south of the White Sea. This area was poorly defended, with LeR 2 fielding just MS.406-equipped LLv 28. It quickly became obvious that this unit was unable to prevent the strafing of troops and installations with its thoroughly obsolete French fighters, so on 15 September the ten Brewsters of Detachment Luukkanen were transferred to Hirvas. The appearance of Model 239s over the isthmus had the desired effect, for the Russians ceased their attacks. The Brewsters had returned to Römpötti by month-end.

The detachment's sole clash occurred on 20 September when Capt Karhunen's swarm engaged ten fighters heading eastwards near Peninsaari. With two Brewsters staying above the enemy formation in order to provide top cover, the remaining pair engaged the 'Spitfires' and quickly shot two of them down – their opponents were actually Yak-7s of 21.IAP, KBF.

Licking their wounds after two months of serious losses at the hands of LeLv 24, the Soviets were conspicuous by their absence in the air over the frontline until 25 October. On this date, four Hurricanes were engaged by the 1st Flight near Tolli lighthouse, and the Finns made up for the recent

Engines idling, a swarm of Brewsters from 1/LeLv 24 prepare to take off on a mission from Römpötti on 6 October 1942. The aircraft closest to the camera is BW-373 'White 3', which was assigned to SSgt Tauno Heinonen. Other fighters visible in this panoramic view include BW-375, BW-393 and BW-382. Three of LeLv 24's four flights operated from Römpötti during the great air battles fought over the eastern Gulf of Finland between 14 and 20 August 1942. During this six-day period, the unit claimed 39 aircraft shot down for the loss of just a single pilot (*SA-kuva*)

lack of activity by downing all of them. The following day, just before noon, Capt Karhunen's flight jumped two Il-4 (formely DB-3F) bombers, escorted by nine fighters, over the coast near Oranienbaum. MSgt Eero Kinnunen (in BW-351) despatched both Ilyushins, while the remaining Brewsters claimed two fighters from the fleeing formation. Thirty minutes later, 1Lt Hans Wind sortied with four Brewsters into the same area and found 15 'Ratas', four of which were duly destroyed.

On 30 October Capt Eino Luukkanen returned to Oranienbaum with his flight, where they engaged a Pe-2 and two escorting 'Spitfires' (possibly Yak-1s). He shot one of the fighters down during his first pass, and as he prepared for a second attack, Luukkanen spotted eight I-16s of 71.IAP, KBF racing to the aid of the Pe-2. A 20-minute melee ensued, during which three Russian aircraft were downed (one of which was an I-16 credited to Luukkanen, flying BW-393) and Sgt Paavo Tolonen was killed when his Brewster (BW-376) fell victim to one of the 'Ratas'.

Within a week of this action taking place, Eino Luukkanen (whose tally now stood at 17 kills) was promoted to major and given command of LeLv 30. His place as 1st Flight commander was duly taken by 1Lt 'Hasse' Wind (who had claimed 14.5 kills to date).

In early November LeLv 24's four flights were instructed to move to LeR 3's Suulajärvi base in the middle of the Karelian Isthmus, and on the 14th and 15th of the month the headquarters, 1st, 3rd and 4th Flights moved in – the 2nd Flight followed suit a week later, coming back south after its extended stay at Tiiksjärvi. For the first time since the beginning of the Continuation War, LeLv 24 was together again on one airfield.

The airspace over the Gulf of Finland remained quiet until 22 November, when, at 0915 hrs, 1Lt Aulis Lumme led six Brewsters into battle against an equal number of Yak-7s from 21.IAP, KBF. Three Russian fighters were claimed to have been destroyed during the 25-minute dogfight that took place west of Kronstadt. A lone Il-2 which

BW-352 'White 2' of 2/LeLv 24 is seen during a visit to Hirvas on 7 August 1942. The fin sports the silhouettes of MSgt Eero Kinnunen's 12.5 Continuation War victories up to that point in the conflict, all but one of which had been claimed in this aeroplane. The fighter also boasts the 2nd Flight's 'Farting Elk' emblem just aft of the fin leading edge, this marking being inspired by a character in the Walt Disney production of *Hiawatha*. Kinnunen was shot down and killed by flak in this aircraft on 21 April 1943, by which time he had increased his tally to 22.5 victories (15 downed in BW-352). Returning to the 2nd Flight's unique 'Farting Elk' marking, this was adopted following its posting to the wilderness of Tiiksjärvi in early January 1942. Painted in black, the small emblem adorned the fin of most of the flight's Brewsters from May 1942 through to 15 February 1943, when 2/LeLv 24 was reformed following its absorption of the 4th Flight. The flight's groundcrew would take great pleasure in 'zapping' aircraft visiting Tiiksjärvi with the 'farting elk' badge, and a Blenheim, a Dornier Do 17 and an Ilyushin DB-3 are all known to have been adorned with this emblem!

strayed into the midst of the engagement was also quickly despatched. Following the rearming and refuelling of its Brewsters, 1Lt Lumme's division was sent aloft once again, together with Capt Karhunen's flight. Heading back to Oranienbaum, the Finns bounced several I-16s that were escorting an Il-4, and all three fighters from 21.IAP, KBF, as well as the bomber, were destroyed.

The final clashes of the year took place 24 hours later. The first flight to see action was 4/LeLv 24, which intercepted six Pe-2s, escorted by five Tomahawks, at 1100 hrs between Lavansaari and Seiskari. Two bombers and two fighters were duly claimed to have been destroyed. Thirty minutes later the 1st Flight attacked three Pe-2s and their I-16 escorts. While the latter had their hands full fending off the rest of the flight, two Brewsters despatched two of the bombers – a solitary 'Rata' was also destroyed. The morning came to a close with yet more fighting for LeLv 24, and this time it was the 3rd Flight's turn to engage the enemy. Eight Tomahawks were chased back over the Russian frontline, with their pilots' attempting to escape the Brewsters by sheltering behind a flak barrage thrown up by the numerous anti-aircraft batteries surrounding Oranienbaum. All but one succeeded, with the sole victim falling at 1230 hrs to WO Juutilainen in BW-351. This was the 34th, and last, victory scored by Finland's 'ace of aces' in a Brewster Model 239.

MSgt 'Lekkeri' Kinnunen sits perched on the tailplane of BW-352 at Tiiksjärvi in June 1942. This photograph was taken to mark the occasion of his final kill with the 2nd Flight (a Hurricane), which he claimed on the 8th of that month. By the time he scored his next victory, on 31 August, Kinnunen had transferred to 1/LeLv 24. On 1 December 1942 he became the air force's youngest warrant officer at just 24 years of age

4/LeLv 24's BW-370 'Black 4' soaks up the sun between sorties at Römpötti in August 1942. This immaculate-looking fighter was regularly flown by deputy flight leader 2Lt Aulis Lumme, who claimed 16.5 victories (4.5 in BW-370) during the course of 287 sorties. Aircraft assigned to 4/LeLv 24 initially wore only the flight's osprey emblem on the fin, but most of its Brewster were also subsequently adorned with the squadron's lynx motif on the forward fuselage. The osprey had first appeared on the fin of a Model 239 in late April 1942, and other aircraft were similarly marked as the year progressed. This badge was chosen due to the flight's location at Hirvas, in the middle of the Karelian wilderness – an area that abounded with birds of prey, including the mighty osprey

Sgt Erik Lyly poses in front of BW-370 at Römpötti in August 1942. A member of the 3rd Flight from 19 January 1942, Lyly scored his two Brewster victories flying 1/LeLv 24's BW-374 (lent to the 3rd Flight – see photograph on page 50) in July-August 1942. On 8 March 1943 he was posted to LeLv 34, where he claimed a further six victories with the Bf 109G. Lyly had completed 188 sorties by war's end

Amongst the squadron 'hacks' assigned to LeLv 24 during the Continuation War was veteran de Havilland DH 60X Moth MO-103, which is seen at Hirvas just prior to its untimely demise on 10 July 1942. Experiencing engine failure in flight over a thick forest, future 29.5-kill ace Sgt Emil 'Emppu' Vesa somehow managed to stall the biplane in at a very low speed. And although the pilot emerged with little more than cuts and bruises, MO-103 was a write-off

No encounters with Russian aircraft were reported in December, and LeLv 24 commenced the New Year with 23 serviceable Brewsters. January also proved to be quiet, and in February a number of the unit's most successful pilots were transferred to the newly-formed LeLv 34, which was to be equipped with the first Messerschmitt Bf 109Gs that were scheduled to arrive from Germany in March.

This unit was intended to be a 'crack outfit' from the start, and despite protests from squadron commanders, its CO, Maj Erkki Ehrnrooth, was permitted to hand-pick pilots from any Finnish fighter unit. Amongst those chosen from LeLv 24 were WO Ilmari Juutilainen (36 kills), 1Lt Lauri Pekuri (12.5 kills) and MSgt Eino Peltola (7.5 kills).

Aside from losing pilots to LeLv 34, LeLv 24 had also suffered such attrition in battle during 1942 that it was left with insufficient aircraft to man four flights, so the unit was reorganised as follows;

Lentolaivue 24 on 11 February 1943

Commander	Lt Col Gustaf Magnusson, with HQ at Suulajärvi
1st Flight	Capt Jorma Sarvanto at Suulajärvi with eight Brewsters
2nd Flight	1Lt Iikka Törrönen at Suulajärvi with eight Brewsters
3rd Flight	Capt Jorma Karhunen at Suulajärvi with eight Brewsters

From August 1942 onwards LeLv 24 fought primarily with units of the Red Banner Baltic Fleet air force over the Gulf of Finland. One of those units regularly encountered was 71.IAP, KBF, whose I-153 'Silver 93' is seen here being connected to a Hucks type starter truck at Lavansaari in August 1942

LeLv 24 saw its first action of 1943 when, at noon on 23 February, Capt Karhunen's flight of six Brewsters bounced four Pe-2s, escorted by twelve I-16s, south of Lavansaari. They returned to base to report the destruction of six fighters, with WO Eero Kinnunen (in BW-352) and Sgt Viljo Kauppinen (in BW-357) both claiming two apiece. Ten days later the Russians lost three I-153s and one I-16 in two clashes in the same area.

On 10 March another large-scale interception took place when seven Pe-2s and ten MiG-3s were spotted heading for Kotka. Eight Brewsters, led by 1Lt Törrönen, attacked the formation over Haapasaari at 1530 hrs, forcing them to turn back. The Finns chased their quarry as far east as Oranienbaum, forcing one bomber and six fighters down onto the ice that covered the Gulf of Finland – only heavy flak prevented the Brewsters from destroying the whole formation.

RED BANNER BALTIC FLEET OFFENSIVE

In an attempt to stop Soviet submarines entering the Baltic Sea come the spring thaw in 1943, the Germans had used the cover of the winter weather to lay a double anti-submarine net across the Gulf of Finland from Porkkala to Naissaari, in Estonia. As a further anti-submarine measure, a double mine belt had also been sown at the same time further east between Kotka and Narva. In order for these protective boundaries to be kept operable, men and equipment had to be shuttled between the island 'links in the chain', with servicing vessels based at a supply base at Kotka.

For over a year these supply points, and bases along the belt, became the primary targets of the Red Banner Baltic Fleet air force. At around this time the air arm also started exchanging its I-153s and I-16s for La-5s and Yak-1s and -7s, as well as increasing the numbers of Pe-2s and Il-2s.

Pilots from 2/LeLv 24 crowd both on and around SSgt Eino Peltola's BW-356 at Tiiksjärvi on 26 May 1942. They are, from left to right, on the ground, Sgt Sulo Lehtiö and Sgt Oiva Lehtinen (1.5 victories), sitting on the wing, 1Lt Väinö Pokela (5 victories) and 1Lt Lauri Pekuri (18.5 victories), standing in the back row, Sgt Paavo Koskela (3 victories), MSgt Eero Kinnunen (22.5 victories) and SSgt Heimo Lampi (13.5 victories), and sitting on the engine cowling, SSgt Eino Peltola (10.5 victories) and Sgt Urho Lehto (3 victories) (*SA-kuva*)

Better aircraft, allied with better tactical training, made the Russians more dangerous opponents.

The Soviet offensive commenced as soon as the sea was free of ice, and LeLv 24 was given the responsibility of protecting the belt, and its supply points. The first action of the spring occurred on 14 April, when 1Lt Wind's four Brewsters were pitted against 30 Yak-1s and La-5s that were escorting a westbound bomber formation over the eastern reaches of the Gulf of Finland. Undaunted by the odds, the Finns fought a valiant low-level dogfight for over 30 minutes, claiming five fighters without loss.

Four days later the first major air battle of the campaign was played out. 1Lt Aulis Lumme (in BW-370) scrambled with seven Brewsters from Suulajärvi at 1700 hrs, followed five minutes later by 1Lt Joel Savonen (in BW-375) and a further six Model 239s. The fighters were vectored onto eight Il-2s of 7.GShAP, KBF and fifty fighter escorts from 21.IAP, KBF detected west of Kronstadt. Following an hour-long combat, the Finns emerged unscathed, with claims of two Il-2s and 18 fighters shot down.

On the morning of 21 April all three Brewster flights intercepted 35 Yak-1, LaGG-3 and La-5 fighters in the Seiskari-Kronstadt area. Initially, only 3/LeLv 24 had been scrambled, its six fighters being led into action by Capt Karhunen. Capt Törrönen's six-strong 2nd Flight took off shortly afterwards when the strength of the opposition became known, and Capt Sarvanto arrived with five Brewsters from the 1st Flight soon after the interception had been effected.

Battling against overwhelming odds, two Finnish pilots were killed, SSgt Tauno Heinonen (in BW-354) being downed by an La-5 of 4.GIAP,

This time it is the turn of 3/LeLv 24's pilots to line up for a group shot, which was taken in front of Capt Jorma Karhunen's BW-366 at Römpötti on 10 September 1942. They are, from left to right, Sgt Eero Pakarinen (3 victories), 2Lt Jalo Ahlsten, 1Lt Martti Salovaara (3 victories), flight leader Capt Jorma Karhunen (31 victories) and his dog 'Peggy Brown', WO Ilmari Juutilainen (94 victories), SSgt Jouko Huotari (17.5 victories) and Sgt Erik Lyly (8 victories). Note that Pakarinen, Ahlsten and Salovaara are all wearing life vests, denoting that they are part of the alert swarm (*SA-kuva*)

A swarm of Brewsters from 2/LeLv 24 are seen on patrol from Tiiksjärvi in late September 1942. This photograph was taken by flight leader Capt Pauli Ervi, flying BW-384. Nearest to the camera is SSgt Lampi's BW-354, followed by MSgt Kinnunen's BW-352, WO Turkka's BW-357 and finally SSgt Peltola's BW-356. All four pilots appear in the official listing of Finnish aces, and all of them scored kills in these very aircraft during the Continuation War. Only Heimo Lampi and Yrjö Turkka would survive through to war's end, however

KBF over Oranienbaum, and Winter War veteran, and 22.5-kill ace, WO Eero Kinnunen (in BW-352) being hit by flak near to where his compatriot was shot down. The Russians paid a high price for this success, however, as the remaining pilots claimed 19 aircraft destroyed – 4.GIAP and 21.IAP, both from the Red Banner Baltic Fleet, are known to have suffered casualties.

On 2 May the Red Banner Baltic Fleet air force attacked Kotka, and 18 Brewsters engaged 30 LaGG-3s of 3.GIAP, KBF south of the target. In an hour-long combat which raged across the Gulf of Finland, the 2nd Flight managed to send four fighters crashing into the sea, although at some cost. Flight commander, Capt Iikka Törrönen (yet another Winter War veteran, and an 11.25-kill ace) was reported missing in BW-380. 1Lt Lumme assumed command of the flight.

Forty-eight hours later twelve Brewsters attacked five Il-2s that had ten I-153s flying as close escort and a dozen LaGG-3s and Tomahawks performing the top cover role. Despite having both a tactical and numerical advantage, the Russians lost nine aeroplanes to the Finns' one – Sgt

Having climbed through the solid cloud base below them, SSgt Lampi and MSgt Kinnunen have swapped places for this shot, which was taken by Capt Ervi later in the sortie. WO Turkka has remained in position, however, although SSgt Peltola appears to have dropped out of the viewfinder – perhaps he is still attempting to find his way out of the undercast! On 21 November 1942 2/LeLv 24's five remaining Brewsters at Tiiksjärvi headed back south, the flight's 'temporary' posting at last having come to an end after almost 11 months

Jouko Lilja was killed when he was jumped (in BW-388) by LaGG-3s of 3.GIAP, KBF.

Of the nine communist aircraft to be destroyed, 1Lt 'Hasse' Wind claimed four of them in BW-393;

'Flying as a wingman to Capt Karhunen, I spotted four I-153s and five Il-2s close to Peninsaari. I commenced my attack by latching onto the tail of an I-153 and chasing it down to sea level. The fighter crashed into the water starboard wing first.

'Whilst we tangled with the fighters, the Ilyushins continued to head towards Kreivinlahti. Determined to stop them, I chased the assault aircraft down, leaving my squadronmates to keep the escorts occupied. The first Il-2 that I attacked was hit in the port wing root, and it caught fire and dived into the sea. My second Ilyushin was downed in exactly the same way.

'I fired at a third as we approached the coast at Shepelevskij, and it began to trail smoke. Burning fiercely, the Il-2 lost height and crashed into the sea near the Tolli lighthouse.

On 1 November 1942 Eino Luukkanen was promoted to major and given command of LeLv 30, which was co-located with LeLv 24 at Römpötti. To mark this occasion, Luukkanen allowed the fin of BW-393 to be adorned with the labels from 17 *Lahden Erikois I* beer bottles. Each one denoted a victory, scored either in the Winter or Continuation Wars. Unfortunately, there is no record of whether Luukkanen consumed the contents of all 17 bottles, or whether his squadronmates helped him with this task! Made CO of LeLv 34 on 29 March 1943, Eino Luukkanen finished the war with 56 victories (from 441 sorties) and a Mannerheim Cross

1/LeLv 24's alert swarm discuss tactics whilst waiting to be scrambled from Römpötti in September 1942. They are, from left to right, SSgt Matti Pellinen (1.5 victories), 1Lt Kai Metsola (10.5 victories), flight leader Capt Eino Luukkanen (56 victories) and MSgt Eero Kinnunen (22.5 victories). The fighter visible to the left of Pellinen is BW-374 'White 4' (*V Lakio*)

'The only evasive action that the Ilyushins had taken was to slide sideways. They rapidly caught fire when hit in the wing root.'

Wind's Il-2s belonged to 7.GShAP, KBF, and they were all seen to crash into the sea. This haul raised his score to 25, leaving him just one kill behind the Continuation War tally of his flight leader, Capt Karhunen. The latter pilot had scored his 31st, and last, aerial victory on this mission.

On 9 May the Russians bombed military installations at Suursaari, and during their return flight to base, the 30-strong fighter escort was jumped by 15 Brewsters, which claimed one La-5 and two Yak-7s shot down. These kills had special significance for LeLv 24, as one of them pushed the unit's victory tally past the 500 mark.

In preparation for the spring offensive, the Red Banner Baltic Fleet had built a new air base on the island of Seiskari during the early months of

The pilot of 4/LeLv 24's BW-367 'Black 6' opens the throttle as he attempts to taxi away from a snow-covered dispersal area at Suulajärvi in late November 1942. Future 19-kill ace 1Lt Erik Teromaa scored four victories with this fighter between 26 October and 23 November 1942. He was later made flight commander of 2/HLeLv 24, and had flown 225 sorties by war's end *(via P Manninen)*

1943. And on the morning of 20 May, 2 and 3/LeLv 24 fought a series of engagements against small fighter formations over the airfield. Three 'Yak-1s' (actually Yak-7s from 13.KIAP – in this instance, 'K' denotes a unit assigned to the Red Banner fleet) and two LaGG-3s were downed without loss. Having refuelled and rearmed at Suulajärvi, both flights duly took on more Soviet fighters over Seiskari that same afternoon. This time six Lavochkin fighters and a Yak-7 were claimed to have been destroyed, again without loss. 4.GIAP admitted losing 'a number' of La-5s.

In six weeks of near-constant action, a veritable handful of obsolete Brewsters had claimed 81 Soviet aircraft shot down for the loss of four fighters (one of which was destroyed by flak). The key to this success was to attack from above, and Finnish pilots were fully aware of the tactical advantages derived from diving on your enemy. By always being above the Soviet formations, the Brewster pilots could fully exploit their tried and tested 'pendulum' tactics, which allowed them to keep the initiative in combat at all times.

To mark the unit's overwhelming success, Marshall Mannerheim paid LeLv 24 a visit at Suulajärvi on 28 May, where he congratulated both

3/LeLv 24's BW-352 was photographed from a Junkers K 43 transport whilst overflying Suulajärvi, on the Karelian Isthmus, on 21 February 1943. The fighter had only returned to the unit ten days prior to this photograph being taken (hence the pristine winter whitewash), BW-352 having been grounded for over six months undergoing a comprehensive rebuild at Tampere. Reassigned to WO Eero Kinnunen upon its return to 3/LeLv 24, BW-352 was lost to flak over Oranienbaum on 21 April 1943. The shattered fighter took Kinnunen with it, the 22.5-kill ace having completed over 300 sorties by the time he was killed (*Finnish Air Force*)

From early 1943 onwards, the Soviet fighter units along the Finnish border started to exchange their obsolete Polikarpov fighters for modern designs from Lavochkin and Yakovlev. An example of the latter, Yak-7B 'White 34' of 29.GIAP (Guard's fighter aviation regiment) prepares to taxy out at the start of yet another mission from an undisclosed base in the Leningrad area during the spring of 1943. Pilots within LeLv 24 only discovered post-war that 29.GIAP was one of the units that they had frequently engaged during the Continuation War (*via C-F Geust*)

Deputy leader of 1/LeLv 24, 1Lt 'Hasse' Wind, returns to Suulajärvi in BW-393 'White 7' after completing yet another sortie in February 1943. Enjoying a long and prosperous relationship with this particular Brewster, Finland's second highest scoring ace would claim 26.5 kills with it between 9 January 1942 and 28 September 1943. Wind's flight leader, Capt 'Eikka' Luukkanen, also found BW-393 to be a lucky machine, scoring seven victories whilst using it as his personal mount in the autumn of 1942. Indeed, it was only permanently passed on to Wind after Luukkanen was posted away from 1/LeLv 24 in November 1942

Lt Col Magnusson and his squadron. At the same time Mannerheim announced that Magnusson was to be promoted to command LeR 3, and that his position as squadron commander was to be taken by the senior flight leader, Capt Karhunen. Furthermore, 1Lt Wind was to be put in charge of the 3rd Flight.

Having tasted triumph, tragedy struck the unit just days later when, on 5 June, Winter War veteran WO Martti Alho was killed in the crash of BW-392 near Tampere. This particular Brewster had been extensively modifed by engineers at the State Aircraft Factory, who had fitted it with a much lighter wooden wing and exchanged its 950 hp Wright-Cyclone engine for a Soviet-built 1000 hp Shvetsov M-63 engine retrieved from a downed I-153.

Alho had been sent to Tampere in order to collect the fighter and fly it back to Suulajärvi, where frontline pilots were to evaluate it against both standard Model 239s and enemy aircraft. Shortly after take-off, Alho banked the fighter into a steep climbing turn, as was standard practice in a conventional Brewster. However, this machine had had an extra fuel tank fitted within its fuselage, altering its all important centre of gravity. Full of fuel, the heavily-laden fighter had neither the speed nor the power to cope with the steep turn, and it stalled and then crashed back to earth. Alho died instantly. An ace with 15 kills to his name, 24-year-old Martti Alho was the unit's youngest warrant officer at the time of his death.

This minor tangle on the ice occurred at Suulajärvi on 27 March 1943 when BW-371 taxied into BW-356. The damage inflicted was light, and the fighters were repaired on base. Both Brewsters belonged to 1/LeLv 24, with BW-371 'White 1' being assigned to Winter War ace WO Viktor Pyötsiä (he claimed no kills with it). This particular Model 239 was one of a handful fitted with a more powerful Shvetsov M-63 radial engine taken from a downed Soviet I-153. This major modification took place at the State Aircraft Factory at Tampere, and usually saw the aircraft's wing guns removed (as was the case with BW-371) (*V Lakio*)

On 20 October 1942 WO Juutilainen intercepted what he described as a 'markingless' Heinkel He 111 over the Gulf of Finland, and promptly shot it down. His victim was almost certainly an Ilyushin Il-4 (ex-DB-3F) of the Baltic Fleet's 1.GMTAP, KBF (Guard's mine-torpedo aviation regiment), which was known to have been conducting a mission in the area at the time. This example is also from 1.GMTAP, KBF, and is seen taxying out for take off from a base in the Leningrad area in December 1942 (*G F Petrov*)

Twenty-four hours after Alho's death, the 2nd and 3rd Flights intercepted two mixed formations of four Pe-2s, seven to eight Il-2s and ten to fifteen escort fighters near Kronstadt. Attacking from above, the Finnish pilots once again wreaked havoc without suffering any losses – three Yakovlev fighters were destroyed, along with a single Petlyakov, one Ilyushin and a solitary Lavochkin.

On 17 June ten SB bombers performed a surprise low-level attack on Suulajärvi airfield, which caught LeLv 24's entire fighter strength on the ground. Fortunately, only BW-351 was destroyed, the veteran fighter being parked out on readiness when it received a direct hit that left it a burning wreck.

This attack proved to be something of a parting shot by the Red Banner Fleet, for few encounters were subsequently recorded over the Gulf of Finland. And engagements over the sea also became a thing of the past thanks to the new Seiskari air base, which was utilised by fighter units providing protection to Russian bombers attacking the German frontline. On several occasions Brewster pilots ventured aloft in an effort to tempt their Soviet opponents into action, but any aircraft that did scramble remained firmly within the protective cover of their own anti-aircraft artillery – forbidden territory for Finnish fighter pilots.

When 3/LeLv 24's WO Juutilainen was chosen for reassignment to newly-formed LeLv 34 on 8 February 1943, his trusty BW-364 was assigned to 1Lt Martti Salovaara. The latter pilot chose not to remove Juutilainen's 36 kill markings (28 of them scored in this very machine) as a mark of respect to Finland's ranking ace – hence it was photographed at Immola in May still boasting a fin full of victory bars. Parked behind BW-364 is 4/LeLv 24's BW-383 (*Finnish Aviation Museum*)

WO 'Pappa' Turkka (17 victories) of LeLv 34 visits his old CO, Capt 'Joppe' Karhunen, at Suulajärvi in early May 1943 – sat between the two men is the latter pilot's 'Peggy Brown'. Their 'seat' is the tailplane of Karhunen's BW-366 'Orange 6', which seems to have been freshly marked with 29 kill bars (see the photograph on the back cover for colour details). Jorma Karhunen would score his 31st, and last, kill in this fighter on 4 May 1943. This tally was amassed during the course of 350 sorties, flown in the Winter and Continuation Wars (*SA-kuva*)

LeLv 24's pilot strength was boosted on 6 July when newly-promoted 1Lt Lauri Nissinen returned to the unit after having been away for almost a year studying at the cadet school. A Mannerheim Cross winner, and 24-victory ace, he immediately took command of the 1st Flight.

Although at full strength in terms of pilots, LeLv 24 was beginning to struggle in respect to serviceable machinery. By mid-1943 it had just 22 Brewsters on strength, which was exactly half the number that had been delivered in early 1940. Senior officers within LeR 3 were fully aware of this problem, and on 16 July six Bf 109G-2s from LeLv 34 were flown into Suulajärvi to provide top cover for all forthcoming missions.

Fortunately, the rest of July passed quietly, with the only event of note being the awarding of the Mannerheim Cross (number 116) to 1Lt Hans Wind on the 31st. He duly became LeLv 24's fourth knight.

On 1 August 1943 LeLv 34 moved to the newly-completed base at Kymi, just north of Kotka, and thus expanded its area of responsibility to the Viipuri-Oranienbaum sector. This had previously been LeLv 24's traditional 'hunting ground', but with the Brewsters being both old and few in number, LeR 3 thought it prudent to leave this hotly-disputed area to LeLv 34's far newer and more potent Bf 109G-2s. Working to the east of this sector, LeLv 24 now had fewer opportunities to engage the communists.

1/LeLv 24 mechanic Paavo Janhunen sits by the tail of recently-transferred WO Ilmari Juutilainen's BW-364 'Orange 4' at Suulajärvi in April 1943. The red star-embellished 'skull and crossbones' marking above the serial aptly denotes the lethality of its former pilot. Once a member of the Bf 109G-equipped LeLv 34, 'Illu' Juutilainen had increased his tally to a staggering 94 victories from 437 sorties by war's end. This success made him one of only four double Mannerheim Cross winners

During the second half of 1943, the Russians concentrated on bombing the German-held island of Tytärsaari, as well as vessels sailing along the Estonian coast. The Brewster pilots found it difficult to engage these formations, for although routing via Oranienbaum as they had done since mid-1942, the Red Banner Baltic Fleet air force units had learnt their lesson, and now positioned all escorting fighters several thousand feet higher.

Soviet aircraft had also previously suffered at the hands of LeLv 24 when taking off and landing from Oranienbaum, but now constant air combat patrols along the Russian frontline provided protection for vulnerable bombers and ground attack aircraft. This meant that communist aircraft were now effectively out of the reach of the antiquated Brewsters. Luckily for the Finns, LeLv 34's Bf 109Gs suffered from no such restrictions.

Late in the afternoon of 20 August the Brewsters worked in conjunction with the Messerschmitts for the first time, three *Gustavs* providing top cover for sixteen Model 239s which engaged fifteen Lavochkins over Kronstadt. Enjoying numerical superiority for a change, LeLv 24 claimed the destruction of four LaGG-3s and two La-5s.

On 28 August Winter War veteran Capt Jouko Myllymäki was made CO of the 2nd Flight. Three days later 12 Brewsters were scrambled after a squadron of Lavochkin fighters was detected flying between Koivisto and Oranienbaum. The Finns became embroiled in a deadly dogfight that saw Sgt Sulo Lehtiö shot down and killed in BW-356. In return, the Brewster pilots claimed two La-5s and two 'LaGG-3s' (both Yak-7Bs of 13.KIAP) shot down, the latter pair falling to recently-returned 1Lt Lauri Nissinen in BW-373. These were his first kills since 8 June 1942.

The final great air battles to be fought over the eastern Gulf of Finland prior to the onset of winter took place on 23 September 1943. At 1330 hrs four Brewsters of 3/LeLv 24, escorted by four Bf 109Gs from 1/LeLv 34, intercepted 20 fighters from 4.GIAP, KBF in the vicinity of Shepelevskij lighthouse. The Finns claimed three Yakovlevs and five Lavochkins

1Lt 'Hasse' Wind, flight commander of 3/LeLv 24, poses in full flying gear alongside his well-weathered mount, BW-393 'Orange 9', at Suulajärvi on 12 September 1943. The fin has been marked with all 33 of his kills up to that point in the war. His most recent victory had come on 17 July, when he downed a LaGG-3. Exactly one week after this photograph was taken, Wind destroyed an La-5 in this very fighter to boost his tally to 34 – he would claim a further 3.5 victories by the end of September (*SA-kuva*)

Another official shot from the sequence taken at Suulajärvi on 12 September 1943. 1Lt Hans Wind is now seen standing in front of BW-393, flanked by the fighter's senior mechanic, Sgt Pentti Saaristo, and an unnamed armourer, who is holding the crank for the Brewster's inertial starter. When Wind was made commander of 3/LeLv 24 in June 1943, he took BW-393 with him, having the veteran fighter's rudder colours changed to those of the 3rd Flight (*SA-kuva*)

destroyed, with the Brewster pilots downing three. One of those to taste success was SSgt 'Nipa' Katajainen, flying BW-368 (in which he had scored his third kill on 1 August 1941);

'I was part of the swarm led by detachment leader 1Lt Salovaara. Whilst heading in to attack the Yaks, I spotted three La-5s above and to the side of us and I engaged them instead, managing to score hits on one. It started to trail smoke and fled into a cloud.

'I then dived on a Yak-1, firing many bursts at it. The pilot initially did well to evade my fire, but once I had hit him, causing his fighter to trail smoke, he tried to take cover in the clouds. I closed in and fired one final burst at him, and the fighter dived into the sea about two kilometres north of Shepelevskij lighthouse.

'My aircraft suffered no hits.'

Two-and-a-half hours later 1Lt Wind's seven Brewsters attacked fifteen aircraft returning to Seiskari airfield, and his pilots claimed one Il-2 of 7.GShAP, KBF and six Lavochkins of 4.GIAP, KBF destroyed.

1Lt Wind's 3/LeLv 24 was in the thick of it again on 28 September, when four Brewsters attacked four Il-2s and four Yak-1s north of Shepelevskij lighthouse – three of the fighters were shot down.

October proved to be a quiet month for the unit, however, for although Russian aircraft were spotted performing both interception and bomber escort missions, no combats took place.

On 4 November a single Yak was added to the LeLv 24's score, and a week later a dozen Yak-7s were intercepted by four Brewsters that were escorting a Blenheim sent to photograph the frontline on the Karelian Isthmus. The Russians broke off their attack once they spotted the Model 239s, although the Finns still succeeded in shooting one of them down.

Only seven Finnish pilots were killed in combat whilst flying the Brewster, and one of those was SSgt Tauno Heinonen of 1/LeLv 24. Shot up by a 4.GIAP, KBF La-5 over Oranienbaum on 21 April 1943, the Finn succeeded in crash-landing BW-354 behind enemy lines. It is unclear whether he was wounded during combat, or in the subsequent crash – as can be seen, the impact of the force-landing badly damaged the fighter's cockpit. Heinonen's injuries proved fatal, in any case, for he died whilst being transported to a Russian hospital. BW-354 was the first substantially intact Model 239 to fall into the hands of the Baltic Fleet air force, and it evoked much interest (*via C-F Geust*)

The following day four Brewsters chased a quartet of Il-2s, escorted by four fighters from 13.KIAP, KBF, back to Oranienbaum. SSgt Emil Vesa (in BW-393) downed one Yak-7, whilst swarm leader, 1Lt Vilppu Perkko (in BW-366), got another. The latter pilot had himself been hit whilst downing his Yak, his Russian opponent, Lt V I Borodin, succeeding in severing the port wing of Perkko's fighter with a series of cannon strikes. With his fighter in flames, and having already suffered serious burns, Perkko succeeded in bailing out at a height of 500 metres – he was quickly rescued by a Soviet navy motorboat and made a PoW. Official communist literature has recorded that Perkko's demise was caused by a 'Taran' attack performed by Borodin (who was killed), but other Finnish pilots on the scene denied that there was any contact between the two fighters.

The onset of typically poor winter weather in December reduced the opportunities for aerial combat over the Karelian Isthmus, and this hiatus in the action lasted for over two months.

Now reduced to operating just 17 obsolete Brewsters, LeLv 24's flight composition on New Year's Day 1944 was as follows;

***Lentolaivue* 24 on 1 January 1944**

Commander	Maj Jorma Karhunen, with HQ at Suulajärvi
1st Flight	1Lt Lauri Nissinen at Suulajärvi with six Brewsters
2nd Flight	Capt Jouko Myllymäki at Suuläjärvi with six Brewsters
3rd Flight	Capt Hans Wind at Suulajärvi with five Brewsters

On 29 June 1943 the engine of BW-353 'Orange 5' quit just short of the runway at Suulajärvi, forcing Sgt Kosti Koskinen to crash the fighter into the thick forest that surrounded the base. Despite the damage inflicted to the Brewster, its pilot emerged from the wreckage unscathed. Astoundingly, the State Aircraft Factory succeeded in totally rebuilding BW-353! Prior to being all but written off in this incident, the veteran fighter had seen much action whilst assigned to 3/LeLv 24's SSgt 'Jussi' Huotari. Indeed, the 17.5-kill ace scored no less than eight of his victories with it between 19 July 1941 and 14 August 1942

BW-365 'Black 9' of 2/LeLv 24 overflies the Gulf of Finland during the summer of 1943. The pilot at the fighter's controls is almost certainly 36-kill ace 2Lt Olavi Puro, who scored 5.5 victories with Brewsters (1.5 in this aircraft, in June-July 1943). BW-365 was one of those Model 239s fitted with an M-63 engine, and like the other similarly-modified Brewsters, it was rarely flown due to the chronic unreliability of the Soviet powerplant. The white rudder and black number worn by this fighter denotes its assignment to the newly enlarged 2nd Flight, which adopted the colours of the recently disbanded 4th Flight (*U Sarjamo*)

La-5 'Silver 26' of 159.IAP taxies across a snow-covered airfield near Leningrad in early 1944. This particular fighter regiment belonged to 275.IAD (fighter aviation division), which marked its aircraft with silver-painted spinners and rudders. Fighters from this division were frequently encountered by HLeLv 24 during the hard-fought air battles of mid-1944 (*via C-F Geust*)

The State Aircraft Factory designed and built 24 Viima (Draught) primary trainers during the Continuation War, and VI-15 was delivered as a 'hack' to LeLv 24 on 3 June 1943. The trainer was photographed at Luonetjärvi on 26 October 1943 following its transfer to bomber squadron LeLv 46 (*Finnish Air Force*)

On 14 February 1944 all frontline squadrons within the Finnish Air Force received a prefix denoting their specific role, *Lentolaivue* 24, for example, becoming *Hävittäjä Lentolaivue* 24 (fighter aviation squadron, which was abbreviated to HLeLv 24). One week later, the unit's area of operations was increased through the extension of its patrol line west of the Virojoki-Seiskari frontline, whilst in the east, the Brewster pilots remained responsible for protecting the middle of the Karelian Isthmus. HLeLv 24 was now restricted to flying interceptions only due to the Model 239's obsolescence, with all other missions requiring the personal approval of the regiment CO.

On 22 February the unit suffered its first fatality of 1944 when SSgt Kalevi Anttila struck ice-covered waters in the eastern Gulf of Finland whilst on a radio check flight in BW-371. He had become lost in sea fog, and descended to low level in an attempt to find his way back to base.

The improving weather in March signalled an escalation in Soviet aerial activity, and with all communist fighter units in the area now equipped with La-5s or Yak-9s, the Brewster pilots were very much on the back foot. Enjoying both qualitative and quantitative superiority, the Soviet pilots should have dominated their opponents, yet the Finns had experience on their side, preventing losses and allowing HLeLv 24 to claim three victories during the month's engagements.

On 2 April 1944 the unit recorded its final Brewster victory, when aces 1Lt Joel Savonen (in BW-375) and 2Lt Heimo Lampi (in BW-382)

Another 275.IAD regiment was the Curtiss P-40 Warhawk-equipped 191.IAP, to whom P-40M 'Silver 23' belonged. Returning from a sortie over Finnish territory in early 1944, its pilot, 2Lt V A Ryevin, ran out of fuel and was forced to land on the ice at Valkjärvi, on the Karelian Isthmus – both the pilot and his fighter were swiftly captured. The machine (serial number 43-5925) was later repainted in Finnish colours and flown on a series of evaluation flights, serialled KH-51 ('KH' stood for Kittyhawk)

BW-374 'Black 6' of 2/HLeLv 24 is seen at Suulajärvi on 8 May 1944 – the day it was handed over to sister-squadron HLeLv 26. 2Lt Eero Riihikallio had flown this machine from January 1943 until the spring of 1944, using it to down 4.5 of the 6.5 kills he achieved with the Brewster (*SA-kuva*)

bounced two LaGG-3s near the Ino lighthouse. The former claimed one damaged, and Lampi's victim was seen to crash on the ice.

Two days later the first Bf 109G-2s were handed over to the 1st Flight from HLeLv 34 (which had received brand new G-6s), and during the course of the month the remaining two flights also re-equipped. The first five Brewsters to leave HLeLv 24 for HLeLv 26, at Heinjoki, departed on 8 May, and the remaining four followed suit three weeks later.

HLeLv 24 had seen near constant action with fewer and fewer Brewster Model 239s from 25 June 1941 to 26 May 1944. During this period its pilots had claimed 459 Soviet aircraft shot down for the loss of 15 aircraft in combat, four in accidents and two in air raids. Twelve pilots had been killed and two had become prisoners of war. Few fighter squadrons in World War 2 would be able to match HLeLv 24's Continuation War kill-loss ratio of 30.6 victories for every Brewster shot down by the enemy.

LENTOLAIVUE 24's LEADING BREWSTER ACES

Rank	Name	Flight	Victories
1Lt	Wind, Hans*	4, 1, 3	39
WO	Juutilainen, Ilmari*	3	34
Capt	Karhunen, Jorma*	3	26.5
2Lt	Nissinen, Lauri*	3, 2, 1	22.5
WO	Kinnunen, Eero+	2, 3	19
SSgt	Katajainen, Nils	3	17.5
Capt	Luukkanen, Eino	1	14.5
MSgt	Alho, Martti#	4	13.5
1Lt	Pekuri, Lauri	2	12.5
1Lt	Lumme, Aulis	4	11.5
Capt	Törrönen, Iikka+	4, 2	10.5
1Lt	Kokko, Pekka	3	10

* - Mannerheim Cross
\+ - killed in action
\# - killed in flying accident

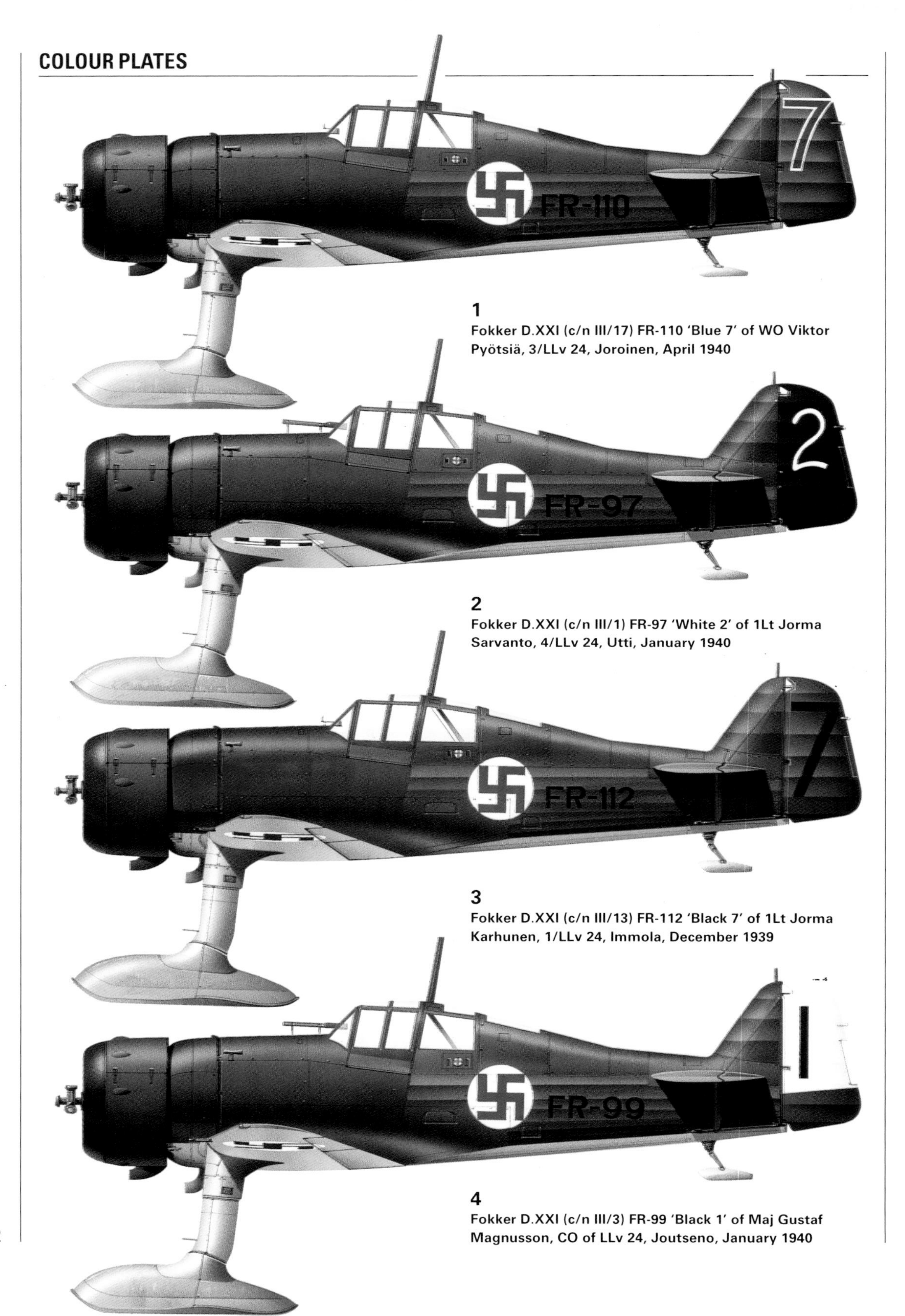

1
Fokker D.XXI (c/n III/17) FR-110 'Blue 7' of WO Viktor Pyötsiä, 3/LLv 24, Joroinen, April 1940

2
Fokker D.XXI (c/n III/1) FR-97 'White 2' of 1Lt Jorma Sarvanto, 4/LLv 24, Utti, January 1940

3
Fokker D.XXI (c/n III/13) FR-112 'Black 7' of 1Lt Jorma Karhunen, 1/LLv 24, Immola, December 1939

4
Fokker D.XXI (c/n III/3) FR-99 'Black 1' of Maj Gustaf Magnusson, CO of LLv 24, Joutseno, January 1940

5

Brewster Model 239 BW-390 'White 0' of 2Lt Kai Metsola, 1/LLv 24, Nurmoila, October 1941

6

Brewster Model 239 BW-357 'White 3' of 1Lt Jorma Sarvanto, 2/LLv 24, Rantasalmi, July 1941

7

Brewster Model 239 BW-368 'Orange 1' of SSgt Nils Katajainen, 3/LLv 24, Kontupohja, March 1942

8

Brewster Model 239 BW-378 'Black 5' of Capt Per-Erik Sovelius, CO of 4/LLv 24, Lunkula, October 1941

9
Brewster Model 239 BW-371 'White 1' of WO Viktor Pyötsiä, 1/LeLv 24, Suulajärvi, March 1943

10
Brewster Model 239 BW-354 'White 6' of SSgt Heimo Lampi, 2/LeLv 24, Tiiksjärvi, September 1942

11
Brewster Model 239 BW-393 'Orange 9' of Capt Hans Wind, CO of 3/HLeLv 24, Suulajärvi, April 1944

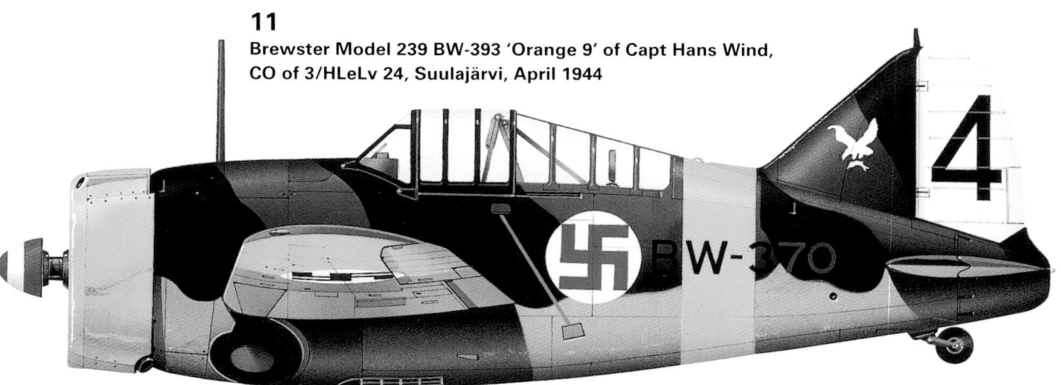

12
Brewster Model 239 BW-370 'Black 4' of 1Lt Aulis Lumme, 4/LeLv 24, Römpötti, October 1942

13
Brewster Model 239 BW-393 'White 7' of 1Lt Hans Wind, CO of 1/LeLv 24, Suulajärvi, January 1943

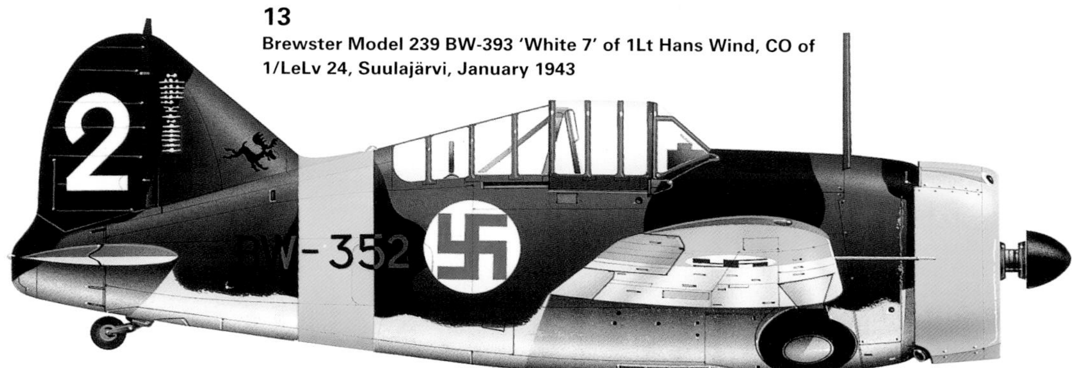

14
Brewster Model 239 BW-352 'White 2' of MSgt Eero Kinnunen, 2/LeLv 24, Tiiksjärvi, September 1942

15
Brewster Model 239 BW-384 'Orange 3' of 2Lt Lauri Nissinen, 2/LeLv 24, Tiiksjärvi, May 1942

16
Brewster Model 239 BW-377/'Black 1' of SSgt Tapio Järvi, 4/LeLv 24, Römpötti, October 1942

17

Brewster Model 239 BW-393 'White 7' of Maj Eino Luukkanen, CO of 1/LeLv 24, Römpötti, November 1942

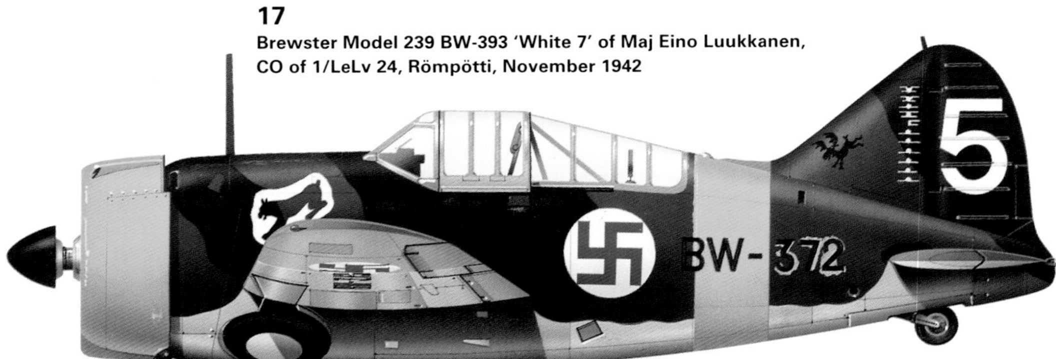

18

Brewster Model 239 BW-372 'White 5' of 1Lt Lauri Pekuri, 2/LeLv 24, Tiiksjärvi, June 1942

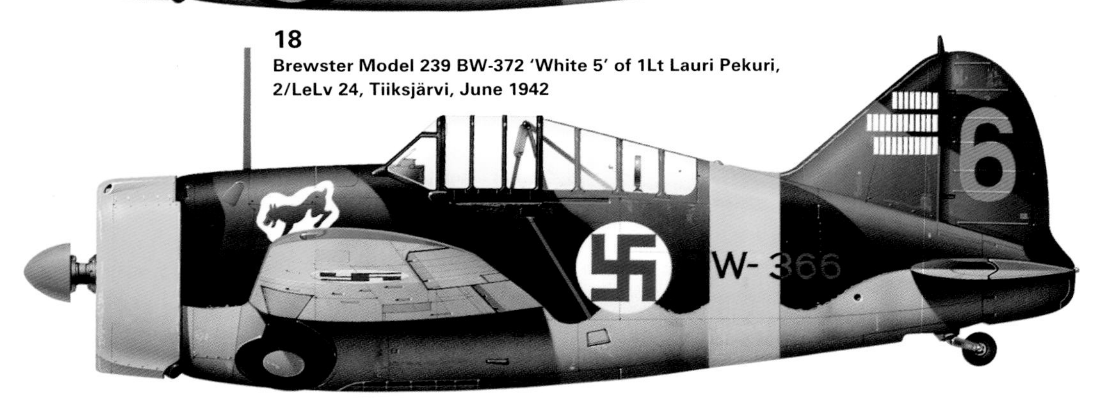

19

Brewster Model 239 BW-366 'Orange 6' of Capt Jorma Karhunen, CO of 3/LeLv 24, Suulajärvi, May 1943

20

Brewster Model 239 BW-386 'Black 3' of MSgt Sakari Ikonen, 4/LLv 24, Kontupohja, April 1942

21
Bf 109G-2 (Wk-Nr 14784) MT-216 'Red 6' of 1Lt Mikko Pasila, 1/HLeLv 24, Suulajärvi, April 1944

22
Bf 109G-2 (Wk-Nr 13393) MT-229 'Yellow 9' of 1Lt Väinö Suhonen, 1/HLeLv 24, Suulajärvi, April 1944

23
Bf 109G-2 (Wk-Nr 10522) MT-221 of 1Lt Jorma Saarinen, 2/HLeLv 24, Suulajärvi, May 1944

24
Bf 109G-2 (Wk-Nr 14754) MT-213 'White 3' of 1Lt Eero Riihikallio, 2/HLeLv 24, Suulajärvi, May 1944

25
Bf 109G-2 (Wk-Nr 10322) MT-231 'Yellow 1' of 1Lt Kai Metsola, 1/HLeLv 24, Lappeenranta, June 1944

26
Bf 109G-6 (Wk-Nr 164929) MT-441 'Yellow 1' of 1Lt Ahti Laitinen, 3/HLeLv 24, Lappeenranta, July 1944

27
Bf 109G-6 (Wk-Nr 164982) MT-456 'Yellow 6' of 2Lt Otso Leskinen, 1/HLeLv 24, Lappeenranta, June 1944

28
Bf 109G-6 (Wk-Nr 165461) MT-476 'Yellow 7' of MSgt Nils Katajainen, 3/HLeLv 24, Lappeenranta, July 1944

29

Bf 109G-2 (Wk-Nr 13577) MT-225 'Yellow 5' of 1Lt Lauri Nissinen, CO of 1/HLeLv 24, Suulajärvi, May 1944

30

Bf 109G-6/R6 (Wk-Nr 165342) MT-461 'Yellow 6' of 1Lt Kyösti Karhila, CO of 3/HLeLv 24, Lappeenranta, July 1944

31

Bf 109G-6 (Wk-Nr 163627) MT-437 'Yellow 9' of SSgt Leo Ahokas, 3/HLeLv 24, Lappeenranta, June 1944

32

Bf 109G-6 (Wk-Nr 167310) MT-504 'Yellow 1' of 1/HLeLv 24, Lappeenranta, September 1944

33
Bf 109G-6/R6 (Wk-Nr 165347) MT-465 'Yellow 7' of 1Lt Atte Nyman, 2/HLeLv 24, Lappeenranta, July 1944

34
Bf 109G-6/R6 (Wk-Nr 165249) MT-477 'Yellow 7' of 1Lt Mikko Pasila, 1/HLeLv 24, Lappeenranta, July 1944

35
Bf 109G-6 (Wk-Nr 165001) MT-460 'Yellow 8' of SSgt Emil Vesa, 3/HLeLv 24, Lappeenranta, July 1944

36
Bf 109G-6 (Wk-Nr 164932) MT-431 of SSgt Pekka Simola, 2/HLeLv 24, Lappeenranta, August 1944

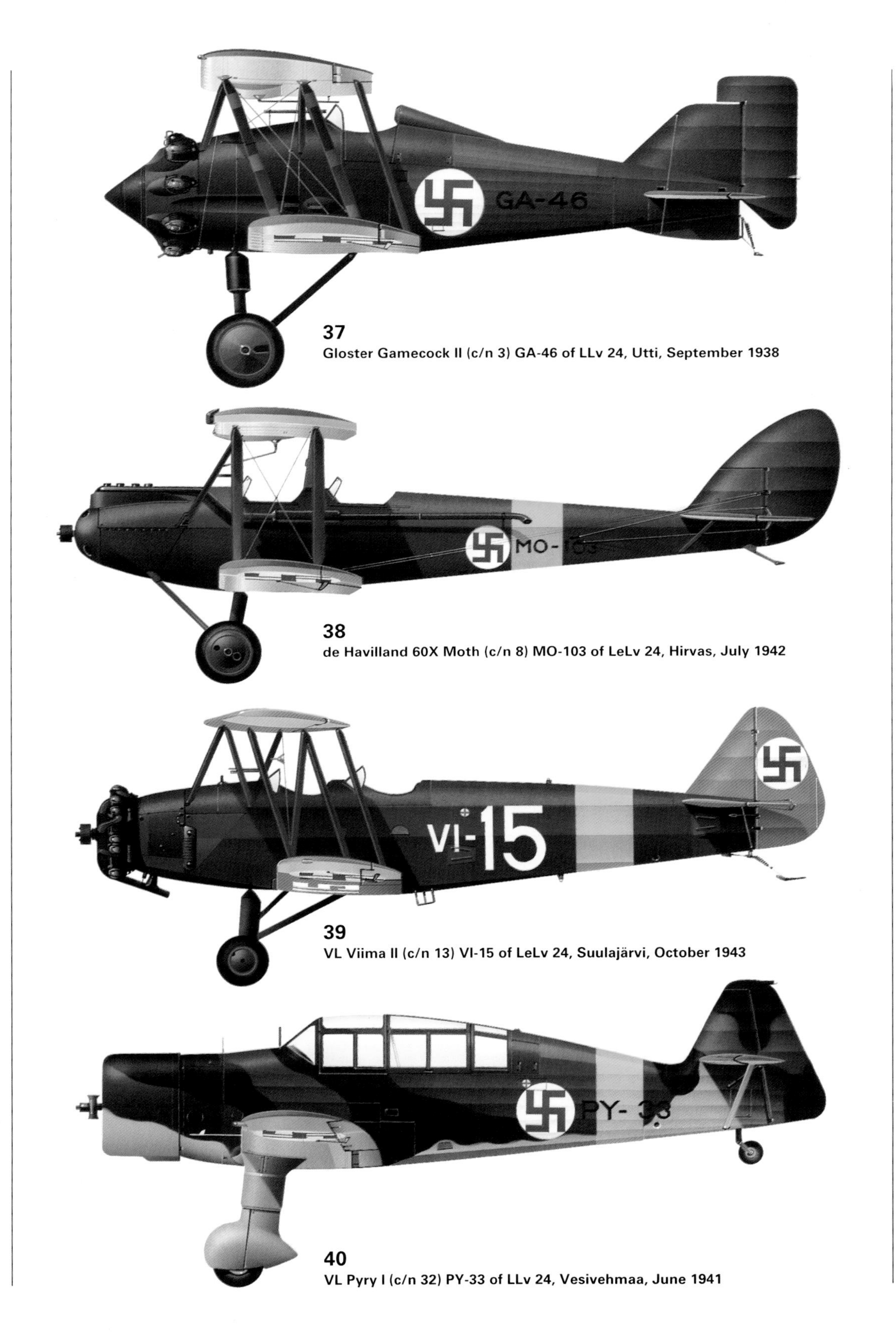

37
Gloster Gamecock II (c/n 3) GA-46 of LLv 24, Utti, September 1938

38
de Havilland 60X Moth (c/n 8) MO-103 of LeLv 24, Hirvas, July 1942

39
VL Viima II (c/n 13) VI-15 of LeLv 24, Suulajärvi, October 1943

40
VL Pyry I (c/n 32) PY-33 of LLv 24, Vesivehmaa, June 1941

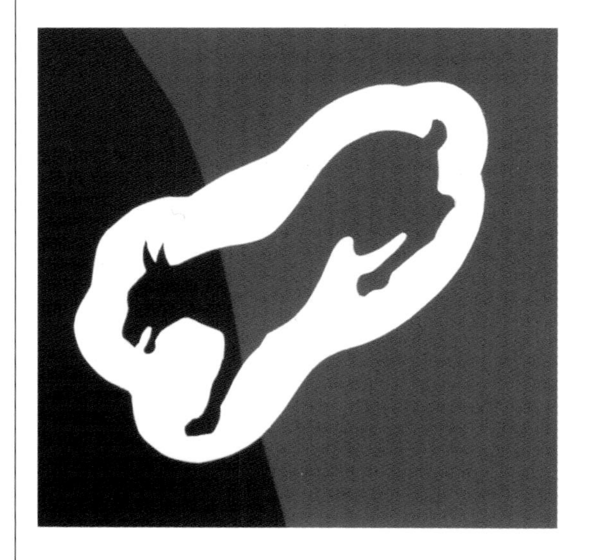

1
LLv 24 unit badge (worn by Brewster Model 239s only)

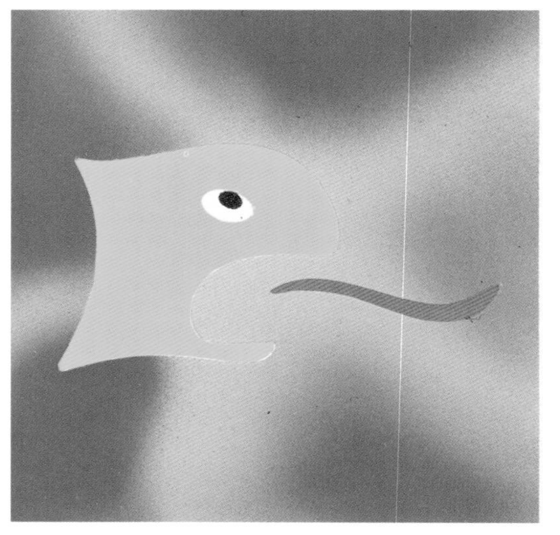

2
HLeLv 24 unit badge (worn by Bf 109Gs only)

3
2/LLv 24's 'Farting Elk' flight badge (worn by Brewster Model 239s only), inspired by Walt Disney's *Hiawatha*

4
4/LLv 24's white osprey flight badge (worn by Brewster Model 239s only)

SOVIET OFFENSIVE

HLeLv 24 commenced transitioning from the battle-weary Brewster Model 239 to the Messerschmitt Bf 109G on 4 April 1944, the 1st Flight having received four G-2s from HLeLv 30 and 34 – the latter units were in turn to re-equip with brand new G-6s. By 11 April the flight was at full strength, and three days later 2/HLeLv 24 took delivery of its first *Gustavs*. This process continued until early May, by which time all three flights had received Bf 109s. The unit's few surviving Brewsters were in turn handed over to HLeLv 26.

Seen at Suulajärvi on 12 May 1944, MT-227 was assigned to 2/HLeLv 24's deputy leader, 1Lt 'Urkki' Sarjamo. He had downed six aircraft with it by 17 June, when the fighter's port wing was shot off by an La-5 and MT-227 crashed into 1st Flight Leader 1Lt Lauri Nissinen's MT-229. Both pilots were killed instantly (*SA-kuva*)

MESSERSCHMITT Bf 109G

After years of being supplied with military aircraft seized by the Luftwaffe, Finland was at last able to buy German-built equipment at the end of 1942. By then, the once all-conquering Nazi war machine had suffered reversals in North Africa that culminated in the encirclement in Tunisia, and stalemate followed ultimately by defeat at the hands of the Soviet Red Army at Stalingrad. The Third Reich badly needed reliable allies, and seeing Finland very much in this vein, Adolf Hitler personally authorised the supply of Messerschmitt Bf 109s. On 1 February 1943 a contract was duly signed covering the supply of 30 Bf 109G-2s to equip one fighter unit.

Sixteen of the aircraft were brand new, priced at 4.0 million Finn marks apiece, whilst the remaining fourteen had been overhauled, and cost 3.6 million. Two batches of aircraft were flown to Finland from Wiener-Neustadt on 13 March, and Erding on 10 May 1943. Their serials were MT-201 to MT-230.

According to the agreement, Finnish losses would be made good, and additional Bf 109G-2s arrived at regular intervals at Pori, on Finland's west coast, via *Feld Luft Park*. However, there was always a delay of two to three months between a fighter being written off and a replacement being flown in. The final example of the 48 eventually supplied arrived on 1 June 1944, and was allocated the serial MT-248.

In early February 1944 the Red Army broke the long-running siege of Leningrad, and in an effort to help the Finnish Air Force contain this new threat, the Germans agreed, on 15 March, to supply sufficient machines to equip another fighter squadron. Within three days the first of 30 new Bf 109G-6s had arrived from Anklam, and by 1 May all these aircraft had been flown in. Serial numbers MT-401 to MT-430 were duly issued.

Following the 9 June 1944 Soviet offensive, the hard-pressed Finnish government asked Germany for immediate military help, and amongst the weaponry it received was a constant flow of new Bf 109G-6s – the first 19 arrived from Insterburg on 19 June. Pilots were kept busy shuttling in new aircraft, and a total of 27 were supplied in June, 19 in July (again from Insterburg) and 24 in August (from Anklam), with the last one arriving on the 30th. They were coded MT-431 to MT-514. Three *Gustavs* failed to reach Finland, MT-473, 474 and 514 being lost en route. Amongst those supplied were two rare G-8s (MT-462 and 483) and a solitary G-6AS (MT-463). The air force acquired no fewer than 162 Bf 109Gs, making it the most common wartime fighter type in Finland.

The Bf 109 was a notoriously difficult aircraft to handle on the ground thanks to its fragile, narrow track undercarriage and heavily framed cockpit, which restricted the pilots' vision when taxying and taking off. Four Messerschmitts were duly written off at Suulajärvi in the first eight weeks after their arrival, the first incident occurring on 12 May when MT-242 collided with MT-236 whilst taking off – the pilot of the latter fighter, 1Lt Martti Salovaara, was killed in the collision. The following day MT-240 was wrecked in another take-off accident, and on 26 May MT-245 suffered an identical fate. Additionally, a further three Bf 109s were so badly damaged that they had to be sent to Tampere to be repaired.

FRIEND OR FOE

On 14 April 1Lt 'Lapra' Nissinen gave the Bf 109G its combat debut with HLeLv 24 when, flying MT-225, he tangled with the enemy;

'Russian aircraft were spotted over the eastern Gulf of Finland, and from my base at Suulajärvi I could see a condensation trail in the sky to the south-east between Lavansaari and Seiskari. Soon, the order came through to scramble, so I took off and headed towards the contrail.

'When I reached an altitude of 500 metres, I observed that the flak batteries at Saarenpää were shooting at the contrailing aeroplane. It immediately changed course and headed in a north-easterly direction, losing its contrail in the process. I followed the aircraft up to a height of 7500 metres, trailing 10 kms behind it. I was slowly gaining on my quarry.

'During the course of my interception, I was continuously radioed information concerning the aircraft's position. It took evasive action over the Gulf of Viipuri, and I lost sight of it. Told to descend, I duly spotted the aircraft two kilometres away, and some 500 metres below me and off to my left. I closed to within 500 metres of its starboard side in an effort to make out its national markings but saw nothing, so I fired a recognition flare and awaited its reply. No response was forthcoming.

1/HLeLv 24 flight leader, and Mannerheim Cross winner, 1Lt Lauri Nissinen poses in MT-225 at Suulajärvi on 4 April 1944. A highly successful fighter pilot who claimed 32.5 kills during the course of 300 sorties, Nissinen was tragically killed when his fighter was struck by the wreckage of 1Lt 'Urkki' Sarjamo's Bf 109G on 17 June. He was the only fighter pilot awarded the Mannerheim Cross to be killed in action (*V Lakio*)

'The twin-engined aircraft was not familiar to me, so I assumed that it was a new Russian type. It was camouflaged in dark shades of paint, and featured no insignia whatsoever – I then remembered from my previous experience with Russian aircraft that they had no upper wing red stars.

'I decided to attack, and from a distance of 300 metres I fired with my cannon and both machine guns, hitting the starboard engine. I reduced my separation distance to just 50 metres and set the starboard engine alight with more cannon rounds. Moments later the aeroplane burst into flames and fell into a steep dive.'

At 1610 hrs a German Junkers Ju 188F-1, coded 4N+NL, of 3./(F) *Aufklärungsgruppe* 22 crashed at Virolaht – the crew bailed out before it hit the ground. The aircraft had taken off from Riga at 1445 hrs to reconnoitre both the ice flow and shipping in the Gulf of Finland, and had entered Finnish airspace without first notifying HLeLv 24. The crew subsequently reported that at 1607 hrs a Soviet fighter (red stars were seen) had shot at them, and fire had been returned. It then disappeared. Two minutes later they spotted the aeroplane that duly shot them down, and its Finnish markings were clearly identified.

The charred wreckage of the Ju 188 was investigated by the Finns, and it was discovered that the aircraft's dirty winter camouflage had obscured

An overall view of 'Lapra' Nissinen's MT-225 at Suulajärvi in April 1944. This particular fighter was the first Bf 109G assigned to the squadron, and it was wrecked in a forced landing on 7 June after it had been shot up by a P-39 from 196.IAP. HLeLv 24's three flights received their new *Gustavs* in sequence, starting with the 1st Flight in mid-April, followed by the 2nd Flight a fortnight later and finally the 3rd Flight in mid-May. The tactical marking 'Yellow 5' on the fighter's nose had been applied by its previous owners, HLeLv 34, which received Bf 109G-6s in place of its G-2s in the early spring of 1944 (*V Lakio*)

On 14 April 1944 1Lt Nissinen took off in MT-225 to intercept an unidentified twin-engined aeroplane detected over the Gulf of Finland. Unable to spot any national markings, and having received no response to identification flares that he had fired, Nissinen shot the bomber down over Virolahti. His victim turned out to be a German Junkers Ju 188F-1 recce aircraft of 3./(F) *Aufklärungsgruppe* 22, coded 4N+NL. This photograph shows well-weathered sister-ship 4N+FL at around the same time (*K Karhila*)

1Lt Mikko Pasila and his mechanics pose in front of 1/HLeLv 24's MT-216 at Suulajärvi. On 18 May Pasila force-landed this aeroplane at Korpela airfield after suffering engine failure. Having joined the squadron on 13 September 1941, Pasila served exclusively with the 1st Flight, completing 200 sorties and claiming ten kills – five with Brewsters and five with Bf 109s (*V Lakio*)

all traces of its national markings. The bomber's yellow eastern front bands had also been of little use, for they could only be seen from below, and had been applied in a pale shade. As a result of this incident, Finnish pilots were immediately issued with recognition manuals that contained photos of all current German types. And Luftwaffe aircraft were forbidden from entering Finnish airspace without prior notice, or clearance from the German air traffic controller at Helsinki-Malmi airport .

On 11 May 1/HLeLv 24 was transferred to Nurmoila, on the Olonets Isthmus, where its Bf 109G-2s were to replace the obsolete Hawk 75As of HLeLv 32 – the latter fighters were proving to be no match for the increasing number of La-5s appearing in the area. The 1st Flight had claimed five Lavochkins in three engagements when the worsening situation on the Karelian Isthmus forced the flight to return to Suulajärvi on 3 June.

HLeLv 24's first encounter with Soviet aircraft following its transition to the Bf 109G occurred on 14 May, when Capt 'Jokke' Myllymäki's swarm attacked two La-5s over Lempaala. The Messerschmitt pilots were escorting two Brewsters sent to photograph the frontline on the Karelian Isthmus when they spotted the fighters, and one La-5 was shot down.

Late May and early June saw Russian troop concentrations north-west of Leningrad substantially increase in size, as did the number of tanks and artillery pieces observed by Finnish fighter reconnaissance aircraft. Unfortunately, senior military figures failed to appreciate the seriousness of these sightings emanating from the fighter recce pilots.

Fighting intensified in the skies over the Karelian Isthmus when Soviet fighters tried to stop the daily recce flights, and between 27 May and 8 June HLeLv 24 engaged the enemy 12 times. Some 21 kills were claimed during this period, but the unit lost 1Lt Heikki Herrala in MT-204 on 2 June, and had 9.5-kill ace SSgt Viljo Kauppinen wounded five days later in MT-225. Herrala had been downed by a Yak-9 from either 14. or 29.GIAP, and Kauppinen had been shot up by an P-39 from 196.IAP.

THE GREAT ATTACK

Following successes on the German front in 1943-44, the Soviet Red Army launched the fourth of its ten strategic attacks on 9 June 1944, and this campaign would ultimately prove to be the only one which failed to reach its goal – the conquest of Finland.

The 'Great Attack', as it became known in Finland, was supported by more than 1300 aircraft from the 13th Air Army, with a further 220 from

1Lt Joel Savonen and 2Lt Heimo Lampi of 1/HLeLv 24 pose with their 'victory sticks' in front of MT-244 at Suulajärvi in May 1944. Neither ace actually scored any kills in this particular fighter, Savonen (who completed 313 sorties and scored eight kills) being assigned MT-235 and Lampi (who flew 268 sorties and claimed 13.5 victotries) MT-232 (*V Lakio*)

the Red Banner Baltic Fleet air force covering the attacking army's left flank. This huge force was tasked with protecting troops on the ground as they advanced across the Gulf of Finland in a 'wedge' no more than 20 km wide at any given point. Further aiding the Soviet attack were the light Nordic summer nights, which permitted round the clock flying.

Opposing this massive air armada on the Karelian Isthmus, Lt Col Gustaf Magnusson's LeR 3 could muster just 16 Bf 109G-6s from HLeLv 34, 18 Brewsters from HLeLv 26 and 14 Bf 109G-2s from HLeLv 24, the latter unit being organised as follows;

***Hävittäjä Lentolaivue* 24 on 9 June 1944**

Commander	Maj Jorma Karhunen, with HQ at Suulajärvi
1st Flight	1Lt Lauri Nissinen at Suulajärvi with five Messerschmitts
2nd Flight	Capt Jouko Myllymäki at Suulajärvi with five Messerschmitts
3rd Flight	Capt Hans Wind at Suulajärvi with four Messerschmitts

The attack commenced on the morning of 9 June, and after quickly breaking through the Finnish defensive line, the Red Army had its enemy beating a hasty retreat. Within ten days the invasion spearhead had crossed the Karelian Isthmus and was on the outskirts of Viipuri, and following the capture of the city on 20 June, the advance was stopped so that the communists could consolidate their newly-won territorial gains .

On the first day of the offensive the Red Air Forces had flown 1150 sorties, leaving their Finnish counterparts overwhelmed by the ferocity of the attacks. Bf 109s were sent aloft in pairs from 0615 hrs to keep a visual check on the Soviet advance, and in three clashes with formations of 100+ aircraft, the Finns claimed one Il-4 and four fighters shot down.

The following day a further 800 sorties were completed, and HLeLv 24 succeeded in destroying 11 aircraft during the course of two morning patrols. The 2nd Flight's deputy leader, 1Lt Urho Sarjamo, was the most successful pilot on this day, claiming one Pe-2 and two La-5s in an early morning sortie, and then engaging the enemy again (in MT-227) whilst patrolling between 0955 and 1100 hrs. His combat report for the latter mission read;

'It was difficult to establish the exact location where the battle started due to poor weather conditions. My swarm engaged 12 Pe-2s east of Kivennapa at a height of 600 metres. We chased them as far as Lempaalanjärvi, the Pe-2s gradually losing height as the pursuit progressed until my target eventually crashed into a forest after crossing a small lake.

'The swarm had split up following this engagement, and I was bounced on my way back to base by a lone Airacobra, which started a turning fight in the vicinity of Riihiö. After a few turns at heights up to 500 metres above the ground, I succeeded in getting on the tail of my opponent, who tried to dive away. I fired at him at distances between 200 and 40 metres, and following a long burst, panels flew off the fighter, it started to trail thick black smoke and eventually crashed into a wood.

'My fighter was struck by a single 12.7 mm round, which passed through the radio receiver and was stopped by the fuel tank armour plate.'

RETREAT

On 11 June HLeLv 24 evacuated its 14 Bf 109Gs from Suulajärvi and headed for Immola. The bad weather that had coincided with the unit's hasty departure from the Karelian Isthmus continued through to the 13th, when only one mission was flown – Capt Wind's six fighters intercepted 30 bombers and 20 fighter escorts between Kämärä and Kaukjärvi whilst carrying out the morning patrol. One P-39 and seven Pe-2s were duly claimed to have been destroyed, with four of the bombers falling to Wind in MT-201.

On 14 June, pilots performing the now routine reconnaissance flight (undertaken by a pair of Messerschmitts) were twice forced to fight their way through enemy aircraft in order to complete their mission. Six Russian aircraft were downed in the process. 1Lt Olavi Puro (in MT-246)

WO Viktor 'Isä-Vikki' ('Father-Vikki') Pyötsiä was a true veteran of the Finnish fighter force, having seen action in both the Winter and Continuation Wars with *Lentolaivue* 24. He is seen here whilst serving with the 1st Flight at Suulajärvi in April 1944. Although assigned MT-244, he was actually shot down in MT-235, which was struck by return fire from an Il-2 on 3 July 1944 – he destroyed the *Stormovik* prior to bailing out of his severely damaged *Gustav*. Pyötsiä was knocked unconscious upon reaching the ground at Nuijamaa, and he was hospitalised for the remainder of the Soviet offensive. 'Isä-Vikki' had flown no less 437 sorties, during which he had scored 19.5 victories (*V Lakio*)

Prior to receiving their own Messerschmitts, the pilots of HLeLv 24 trained with fighters supplied by 1/HLeLv 34, which was also based at Suulajärvi. In one of the first flights attempted by a pilot from HLeLv 24 (on 21 March 1944), Brewster ace MSgt Nils Katajainen was caught out by the Bf 109's tendency to swing on take off. He attempted to correct this by pulling back on the control column, thus becoming airborne, but MT-239 simply stalled port wing down and cartwheeled to a stop. Katajainen was lucky to escape this high-speed crash with just minor injuries (*E Laiho*)

led both of the morning recce flights, and on the second sortie he encountered the largest formation of enemy aircraft that he had ever seen;

'While approaching the target, I observed that a huge bomber formation was releasing its ordnance on the Finnish frontline. I immediately dived at the enemy aircraft, which consisted of approximately 100 bombers and a similar number of fighters. Both 1Lt Saarinen and I were engaged by Russian fighters.

'I succeeded in shooting down two La-5s and one Il-2, all of which fell near the frontline between Vammeljoki and Mustamäki. When the bomber formations finally turned back, we took the opportunity to finish off our reconnaissance tasking.

'Soon after the first wave of bombers had disappeared, a second formation of 70 aircraft (split into three groups) appeared from the east. Our own anti-aircraft artillery kept up a constant barrage against the Russian aircraft as they approached the frontline, and I added to their spirited attack by expending my remaining ammunition on several Pe-2s and a couple of fighters.

'By flying into the midst of such large formations, we successfully confused the gunners aboard the Russian bombers, for they had great difficulty in distinguishing two Messerschmitts from their own fighters. Indeed, their rear gunners did not once open fire on us.'

On 15 June HLeLv 24 moved once again, transferring this time further south-west to Lappeenranta, where the unit would remain until war's end. HLeLv 24 spent the first 24 hours at its new home servicing its fighters.

1/HLeLv 24's Bf 109G-2 MT-229 is seen on a sunny spring day at Suulajärvi in May 1944. This aircraft was assigned to 1Lt Väinö Suhonen, who scored 19.5 victories (two in MT-229) during 261 sorties. On 17 June his flight leader, 1Lt 'Lapra' Nissinen, was flying this very machine when he was hit by the falling wreckage of 1Lt Urho Sarjamo's MT-227. The 32.5 kill ace had also claimed two kills with MT-229 prior to his untimely death (*V Lakio*)

HLeLv 24's long-time home at Suulajärvi boasted numerous facilities not usually found at other (more temporary) sites occupied by the unit during World War 2. Amongst the most important amenities were the many blast pens that had been dug into the earth, and then lined with logs felled from nearby forests. This particular pen has been covered by camouflaged netting, under which is sat MT-231 of 1/HLeLv 24. This aircraft was assigned to 10.5-kill ace 1Lt 'Kaius' Metsola, who served with the flight throughout the Continuation War. He claimed a solitary victory with MT-231 (an Il-2 on 17 June 1944) (*V Lakio*)

The unit's first full day of combat from Lappeenranta proved to be a bad one. At 0620 hrs 1Lt Lauri Nissinen (in MT-229) led ten Messerschmitts up to intercept a massive formation of bombers, assault aircraft and fighters that had been spotted at a height of 2000 metres heading for the Finnish positions between Kaukjärvi and Perkjärvi. 2Lt Heimo Lampi (in MT-235) was flying as Nissinen's wingman;

'We extract the very last horsepower out of our engines as we struggle to gain both enough altitude and speed for our first diving attack. I hear Nissinen and Sarjamo agree that the latter's division should attack the enemy aircraft directly ahead of us, while Nissinen's division tries to gain further altitude in order to hit the enemy formation from above.

'We continue climbing until it is time to push over into a dive and head down into the Russian formation. Now it is our turn to attack our old acquaintances, the Lavochkin La-5s. I bank left into a shallow dive and pick out my targets – a section of four fighters, which will soon be forming a "carousel" as they follow each other around in circles in an attempt to shoot me down.

'As I am just about to commit myself to the attack, Nissinen's Messerschmitt appears next to me, and he waves at me to follow him – we cannot communicate by radio due to the incessant chatter synonymous with combat. The remaining members of our swarm fail to spot Nissinen

A gathering of aces. Clutching his victory stick, unit CO Maj Jorma Karhunen (31 victories) smiles for the camera in front of 2/HLeLv 24's MT-213 at Suulajärvi in May 1944. The other pilots in this photograph are, from left to right, SSgt Tapio Järvi (28.5 kills), 1Lt Lauri Nissinen (32.5 kills), 1Lt Jorma Saarinen (23 kills), Sgt Arvo Koskelainen (5 kills), SSgt Paavo Koskela (3 kills) and 1Lt Aulis Lumme (16.5 kills). Bf 109G-2 MT-213 was the personal mount of 16.5-kill ace 1Lt Eero Riihikallio, who claimed three kills with it

Watching mechanic Lauri Kitinoja turn the inertial starter crank of his fighter's Daimler-Benz DB 605A engine, 2/HLeLv 24's 1Lt 'Jotte' Saarinen prepares to fire up MT-221 at Suulajärvi on 12 May 1944. Having scored his first two Messerschmitt kills with this aircraft, the 23-victory ace eventually moved on to Bf 109G-6s in late June, being allocated MT-452 and then MT-478. Saarinen was killed on his 139th sortie (on 18 July) when the latter fighter struck an embankment during an attempted forced-landing – the *Gustav* had been shot up by an La-5 from 159.IAP (*SA-kuva*)

and I breaking away from the attack, and they continue diving down towards the enemy. Nissinen nods his head and goes into a steep dive. Puzzled by his actions, I follow him – he is the most aggressive combat pilot in the unit, yet he flies straight past the Russian formation and dives into cloud below them. I later learned that he had received a radio message from our ground controller telling him to break off his interception and attack a large formation of unescorted assault aircraft near Viipuri.

'The layer of cloud is only 100 metres thick, and we pass through it in tight formation at 600 km/h, before levelling off some 50 metres below the overcast. I can hear from my radio that a ferocious combat is taking place above the clouds, and I also realise that an enemy fighter could pop out of the cloud behind us at any moment and shoot us down.

'Keeping an eye on my tail, I am suddenly startled by the appearance of a Messerschmitt fighter as it drops out of the cloud immediately above my leader. Minus its port wing, the aircraft is dropping like a stone. Before I can take any evasive action, the fighter smashes squarely into the middle of Nissinen's aeroplane, and both Bf 109s disintegrate into thousands of pieces. Two black lumps – the burning engines of the fighters – fall to earth ahead of the wings, tails and other parts of the shattered fuselages. I am stunned by this collision, which has occurred just 20 metres away. Through sheer instinct, I push my machine into a dive, hoping to spot one of the pilots in a parachute. I am perfectly aware that no one could have survived such a collision, yet I must keep on watching.'

The fighter that struck Lauri Nissinen's MT-229 was none other than MT-227, flown by fellow ace 'Urkki' Sarjamo. The latter's machine had had its port wing shot off by an La-5 of 159.IAP. The remaining pilots had destroyed four Il-2s and four La-5s during the course of the mission, but these failed to offset the loss of two veteran aces. 1Lt Joel Savonen duly became the acting flight leader of 1/HLeLv 24 in place of Lauri Nissinen.

On 18 June the unit learnt that long-time member, and flight leader, Maj Eino Luukkanen, who now led sister squadron HLeLv 34, had been awarded Mannerheim Cross number 127.

3/HLeLv 24's 1Lt Martti Salovaara was killed in this accident, which occurred at Suulajärvi on 12 May 1944. Opening the throttle to check that his engine was functioning properly, prior to releasing his brakes and commencing his take-off run, Salovaara (in MT-236 to right) created a tremendous dust cloud. Taxying out behind him was SSgt Leo Ahokas in MT-242, who saw the cloud and assumed that Salovaara was already accelerating down the grass runway. He duly advanced the throttle on his fighter and smashed into Ahokas's still stationary Messerschmitt, MT-242 mounting MT-236 and burying its propeller into the *Gustav's* cockpit. Martti Salovaara was killed instantly. Both aircraft were so badly damaged in the accident that they were subsequently written off (*SA-kuva*)

From 19 June the arrival of new Bf 109G-6s from Germany started to make good the losses suffered by the fighter units up to this point in the battle. A direct beneficiary was 3/HLeLv 24, which at last received its full complement of eight aircraft. The flight in turn passed on its three remaining G-2s to the 2nd Flight.

In the early evening of the 19th, 2/HLeLv 24 downed two La-5s. Shortly afterwards, at 2000 hrs, an 18-fighter formation comprising eight fighters from HLeLv 34 (led by the 3rd Flight's Capt Puhakka) and ten from HLeLv 24 (with Capt Wind in command) intercepted several regiments of Russian aircraft near Viipuri. The Finns destroyed six Pe-2s (from 58.BAP), three Airacobras (from 196.IAP), two Il-4s (836.BAP) and two La-5s (401.IAP) without loss – Capt Wind was credited with a trio of kills flying newly-delivered Bf 109G-6 MT-439.

LOSS OF VIIPURI

The air war reached its peak on 20 June, when Russian troops forced their way into the streets of Viipuri, supported by a massive aerial 'umbrella' of fighters and ground attack aircraft. Before midday, both Messerschmitt squadrons had already been embroiled in three large-scale actions which had seen them claim 35 aircraft shot down. And by the end of the day a further five battles had taken place, with the Finns adding an additional 16 victories to increase their overall tally for the 20th to 51 kills – 31 to HLeLv 24 and 20 to HLeLv 34. Multiple losses were suffered by Yak-9-equipped 14.GIAP, La-5-equipped 159.IAP, Airacobra unit 196.IAP and the Il-2 regiments 943. and 35.ShAP, KBF.

HLeLv 24's top scorers on this day were Capt Hans Wind (in MT-439) and 1st Lt Olavi Puro (in MT-201), both of whom claimed five kills during the course of two sorties, and Sgt Eero Halonen (in MT-241), who downed four aircraft again over two missions.

2/HLeLv 24's MT-213 is run up at Suulajärvi on 12 May 1944. This aircraft was one of just a handful of Bf 109Gs that actually had its Luftwaffe 'greys' oversprayed with the standard Finnish black/green camouflage scheme. All Messerschmitts flown by this unit featured toned down discs for the national insignia – these had been introduced on 12 January 1944 (*SA-kuva*)

HLeLv 24 claimed a further six kills on the 21st and nine on the 22nd, although 2Lt Erkki Nukarinen (in MT-442) was killed on the latter date when he was shot down over Tali by Yak-9s from either 14. or 29.GIAP.

Following the capture of Viipuri, the Red Army re-grouped its troops and headed westward in the direction of Tali and Ihantala, utilising the only track in the region suitable for armoured vehicles. The village of Ihantala proved to be a stumbling block for the communists, with the Finnish Army stopping the Red Army's advance in its tracks. Indeed, the Soviet losses in this area were so great that post-war, the Russian government erected a monument in the village that was dedicated to the memory of the 65,000 soldiers killed there.

Supporting their comrades on the ground, HLeLv 24 continued to exact a heavy toll of the Red Air Force units committed to the invasion of Finland. On 23 June the unit claimed 22 victories whilst completing four missions, Capt Wind again topping the list of victors with four kills in his favourite MT-439 during the sortie flown between 1200 and 1305 hrs;

'I encountered about a dozen La-5s whilst on patrol soon after midday. After firing a short burst at one of them, the fighter exploded in mid-air. My second victim started to smoke after I fired at it from close range, and the La-5 duly crashed into a forest near Säiniö.

'I then observed several formations of Ilyushin bombers heading for Viipuri and I immediately gave chase. Shooting two of them down in flames – one crashed at Viipuri and the other at Liimatta – I then ran out of ammunition and reluctantly returned to base.'

The La-5s had belonged to 11.GIAP and the Il-4s to 113.BAD.

Following 72 hours of near constant action, heavy rainfall kept most aircraft from either side on the ground for 48 hours. The Finns managed just a single reconnaissance mission on both days, and on the 25th 5-kill ace, and 2/HLeLv 24 flight commander, Capt Jouko Myllymäki failed to return from a recce flight in MT-221. With the frontline covered by low

All four of 2/HLeLv 24's future top scorers are seen relaxing between sorties at Suulajärvi on 12 May 1944. They are, from left to right, SSgt Tapio 'Tappi' Järvi (28.5 kills), 1Lt Olavi 'Olli' Puro (36 kills), 1Lt Jorma 'Jotte' Saarinen (23 kills) and 1Lt Eero 'Riihi' Riihikallio (16.5 kills). These pilots had scored 30 victories between them when this photograph was taken, and by the end of the Continuation War their combined tally had risen to 104 (*SA-kuva*)

MSgt 'Jussi' Huotari of 3/HLeLv 24 prepares to squeeze into the cockpit of MT-240 at Suulajärvi on 12 May 1944. He was later assigned G-6 MT-440, and scored five kills with it in late June and early July. Serving with the 3rd Flight throughout the Continuation War, Huotari completed 291 sorties and claimed 17.5 aircraft destroyed (*SA-kuva*)

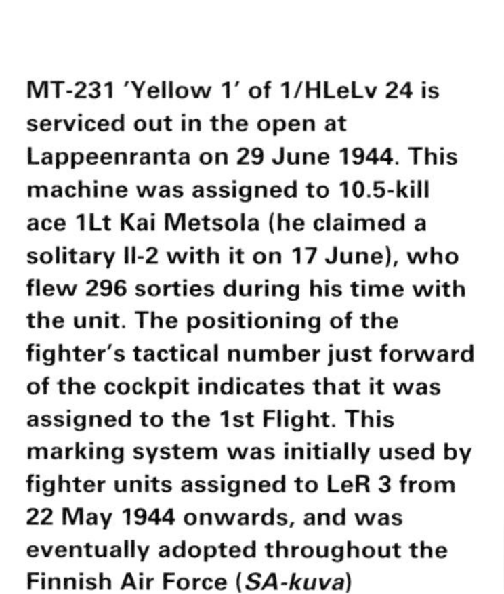

cloud, he had been forced to fly so low in order to carry out his mission that his fighter had clipped trees and struck the ground at high speed. Fellow ace 1Lt Aulis Lumme duly became the acting leader of the 2nd Flight.

An improvement in the weather on 26 June brought an end to the brief hiatus in the action, and before noon the 2nd Flight had intercepted ten Il-2s of 566.ShAP that they had found harassing troops at Tali – half of them were quickly shot down. In the early afternoon 2/HLeLv 24 was again called back to this area by Finnish troops when yet more *Stormoviks* appeared overhead, and this time four Il-2s were destroyed. Later that same day the 3rd Flight engaged Yak-9s of 14.GIAP twice in two hours over Viipuri and Tali. Six Soviet fighters were shot down, five of them falling to Capt Wind in MT-439.

Crowning a great day for the Finnish fighter force, regiment commander Lt Col Gustaf Erik Magnusson was awarded Mannerheim Cross number 129 on the 26th. A most worthy recipient of Finland's highest military award, he had personally developed the fighter force's command and control system in the autumn of 1943. Lacking any effective radar equipment, Magnusson had instead been forced to rely on a series of strategically-placed observation posts which were each equipped with transmitters. Finally, he had staffed the regiment control room at Lappeenranta with seasoned radio intelligence operators skilled in cracking Russian codes. The combination of these various elements was the key to the Finnish fighter force's stunning success against overwhelming odds.

On 28 June HLeLv 24 enjoyed its most successful day of the entire Russo-Finnish War when its pilots claimed 33 kills in five missions – but not without cost. The action

MT-231 'Yellow 1' of 1/HLeLv 24 is serviced out in the open at Lappeenranta on 29 June 1944. This machine was assigned to 10.5-kill ace 1Lt Kai Metsola (he claimed a solitary Il-2 with it on 17 June), who flew 296 sorties during his time with the unit. The positioning of the fighter's tactical number just forward of the cockpit indicates that it was assigned to the 1st Flight. This marking system was initially used by fighter units assigned to LeR 3 from 22 May 1944 onwards, and was eventually adopted throughout the Finnish Air Force (*SA-kuva*)

Bf 109G-6 MT-456 'Yellow 6' of 1/HLeLv 24 has its engine warmed up at Lappeenranta on 29 June 1944. This machine was the mount of 2Lt Otso Leskinen, who claimed his only victory on this very day – he downed a Yak-9 over Noskuanselkä, east of Viipuriin. During the fierce battles of the final Soviet offensive, every flight had four pilots, with their aircraft, standing alert 24 hours a day (*SA-kuva*)

started at 0900 hrs when 11 Messerschmitts, led by Capt Wind, intercepted 20 Pe-2s, 20 Il-2s and several dozen fighters bound for Tali. The Finns downed eight Il-2s, two Pe-2s and an La-5, but 5.5-kill ace Sgt Kosti Keskinummi was wounded when his fighter (MT-437) was shot up over Nuijamaa and he was forced to carry out a belly landing.

Ninety minutes later two Bf 109s took off on a reconnaissance mission specifically generated to locate the spearhead of the Russian attack on the ground. MSgt Nils Katajainen's combat report for this sortie recounts;

'Whilst flying as Capt Wind's wingman on this special recce mission, we were attacked by 20 fighters. I fired at one Airacobra, which burst into flames – Capt Wind saw it burning.

'I broke off my chase of the seemingly doomed P-39 and climbed for altitude. I then spotted another Airacobra and closed to within firing range. Several bursts later, the enemy fighter was ablaze, although I did not see it crash. Nearby, just south of Tali, I noticed an observation balloon, so I dived through the anti-aircraft barrage and shot it down in flames.

'Whilst returning at low-level to friendly territory, I saw seven Il-2s off to my left. I flew straight at them as they passed directly over Tali, and managed to set one of the Ilyushins alight. It hit the ground trailing smoke, and upon coming to a halt the bomber erupted in flames. I fired at a second Il-2, but ran out of ammunition during the course of my attack.

'I saw Capt Wind down one Yak. My MT-436 suffered no damage.'

Wind had in fact destroyed three Yak-9s (which his wingman had mistakenly identified as P-39s) before he was badly wounded when a cannon shell exploded in the cockpit of MT-439. Barely conscious from loss of blood, he just managed to bring his fighter back to base, where, after receiving first aid and while waiting for a transport aeroplane to take him to a military hospital, he filled in a brief combat report;

'En route to our mission objective we were attacked by 20 Yak-9s as we overflew Juustila. Forced to fight for my survival, I had shot two aircraft down and set a third Yak alight when another fighter hit me from the side and behind. Seriously wounded in my left arm, I succeeded in making it back to base, where it took all my strength to perform the landing.'

For Hans Wind, HLeLv 24's top scorer with 75 confirmed victories from 302 sorties, the war was over. Fellow high-scoring ace WO Ilmari Juutilainen of HLeLv 34 had also been cutting swathes through the ranks of the Red Air Forces since the launch of the invasion, and both he and Wind became the first individuals to receive the Mannerheim Cross for a second time on this day – both decorations were unnumbered. They were also the only double recipients during the conflict with the Soviet Union.

Also in action on 28 June was 1/HLeLv 24, which destroyed four Il-2s and two Yak-9s over Tali, while the 2nd and 3rd Flights twice engaged large formations of Russian aircraft in the same area, claiming 12 shot down. One hour after the wounding of Hans Wind, the Bf 109G-6 (MT-438) of 29.5-kill ace SSgt 'Emppu' Vesa was also hit by enemy fire after its pilot had just downed a brace of Il-2s. Struck in the oil cooler, the fighter's engine seized and Vesa was forced to belly land at Juustila. Later in the day 25-kill ace 1Lt 'Olli' Puro was wounded in both legs by shrapnel, although he succeeded in bringing MT-449 back to base. He returned to the unit a fortnight later to score a further 11 victories.

The Soviets are known to have lost over 20 aircraft on 28 June, including Yak-9s from 14. and 29.GIAP, La-5s from 159.IAP, Airacobras from 196.IAP and Il-2s from 448. and 566.ShAP. Three 'Mustangs' were also claimed by HLeLv 24, but as with previous cases of misidentification involving the US fighter, these were almost certainly Yak-9s or Yak-9Ds.

This intense period of fighting continued at 0720 hrs on the 29th, when 1Lt Mikko Pasila (in MT-238) led six fighters on a patrol over Tali. En route, he spotted more than 180 Russian aircraft engaged in a dogfight with 11 Bf 109Gs from 3/HLeLv 34, and he quickly rushed to their aid. Both squadrons jointly destroyed four Pe-2s, two Yak-9s, one La-5 and

Left
Engine cowling hinged open, MT-455 'Yellow 2' of 1/HLeLv 24 is prepared for its next mission at Lappeenranta on 29 June 1944. The aircraft has been parked beneath the fir trees that proliferated at the base. MT-455 was assigned to 5-kill ace Sgt Arvo Koskelainen, who claimed his sole victory in it (a Pe-2) over Kannas on the very day this photograph was taken. Fellow ace 1Lt Kai Metsola (10.5 victories) wrote the fighter off in a landing accident at Lappeenranta on 9 July, the incident being blamed on the pilot's state of total exhaustion (*SA-kuva*)

This Petlyakov Pe-2 bomber was assigned to 12.GPAP, KBF (Guard's dive-bomber aviation regiment) during the summer of 1944. Based in the Leningrad area, the unit suffered heavy losses at the hands of HLeLv 24 during the Soviet offensive, although the primary target for the Messerschmitt pilots during the brief, but bloody, summer campaign was the Il-2, which was the scourge of the Finnish Army (*G F Petrov*)

A common sight on the Karelian Isthmus in the summer of 1944. This Il-2 was the 19th kill for 1Lt Kyösti Karhila, who shot the Ilyushin down at Tienhaara, north-west of Viipuri, on 21 June. It was assigned to 703.ShAP (of 281.ShAD), and its pilot, 1Lt M I Berleyev, was captured by Finnish troops. The fate of his gunner remains unrecorded (*SA-kuva*)

A far less common sight during the summer of 1944. Bf 109G-6 MT-437 'Yellow 9' of 3/HLeLv 24 was damaged in combat on 28 June and forced landed at Nuijamaa by its wounded pilot, 5.5-kill ace Sgt Kosti Keskinummi. The fighter was actually assigned to fellow ace SSgt Leo Ahokas, who scored 12 victories during the course of 189 sorties – his sole kill (an La-5) in this aircraft came on 20 June. Barely visible on the fighter's damaged rudder is the rarely seen yellow lynx's head badge adopted by HLeLv 24 as its official unit emblem on 7 June 1944. This marking resulted from a contest held within LeR 3 in May 1944 to produce new squadron emblems – on 7 June LeR 3's commander, Lt Col Magnusson, accepted a yellow lynx's head for HLeLv 24. This marking remained photographically undocumented until 1975 (*SA-kuva*)

one Il-2 for the loss of HLeLv 24's 1Lt Ahti Laitinen. The 12-kill ace, who had downed four aircraft the previous day, bailed out of MT-439 (previously Hans Wind's mount) when it was shot up by an La-5 from 159.IAP. Landing near Ihantala, Ahti Laitinen was quickly captured – he would be released as part of the truce agreement at Christmas.

The fact that Laitinen was flying Wind's Messerschmitt just 24 hours after the unit's leading ace had been badly wounded in it indicates that the cannon shell had caused more damage to the pilot than to the aircraft. Here, Ahti Laitinen describes his last mission;

'With Sgt Helava flying as my wingman, I headed towards Juustila, where Russian aircraft had been observed. As I overflew the town at a height of 4000 metres, I spotted more than 100 Pe-2 bombers, escorted by more than 50 La-5 and Yak-9 fighters.

'As I dived down to attack a Pe-2 I failed to spot that an La-5 had slipped in behind me and it struck my aircraft with a series of blows from close range. Shrapnel from one of the 20 mm shells that hit the fuselage struck my left leg, and I swiftly nosed the Messerschmitt over into a vertical dive. More shells hit the cockpit and the engine, which caught fire. My damaged fighter was now uncontrollable so I released the canopy, and after several desperate attempts I managed to bail out of the burning aeroplane as it hurtled earthward at 800 km/h. I vacated the cockpit at a height of 2000 metres and slammed hard against the tail, breaking my right arm and leg. My head also struck the aircraft at this point, and I lost consciousness. Quite miraculously, my parachute immediately deployed and I landed still totally unconscious.

'I had come down between the Finnish and Russian frontlines, and the communists got to me first. They dragged me in a tent canvas into their trenches, and after administering first aid, they took me to a field dressing station, where I was interrogated. Once this had been completed, I was loaded into an R-5 and flown to a military hospital in Leningrad. Once my wounds had fully healed, I spent the next six months at various prison camps.'

Following the loss of Hans Wind and Ahti Laitinen within 24 hours of each other, 20-victory ace 1Lt Kyösti Karhila was made acting flight leader of the 3/HLeLv 24.

Airacobra-equipped 196.IAP was yet another fighter regiment assigned to 275.IAD during the summer offensive of 1944. Here, P-39N 'Silver 24' has just been refuelled at its base in the Leningrad area at the height of the campaign to capture Finland. Division markings consisted of a silver-painted tactical number, rudder and spinner (*via C-F Geust*)

The unit's run of high scores continued on 30 June when, between 1045 and 1200 hrs, seven of its fighters combined with eight Bf 109Gs from HLeLv 34 to down 15 aircraft from a force of 200 to 300 that was attacking Finnish positions at Tali and Ihantala. 14.GIAP and 404.IAP are known to have lost Yak-9s, as did Airacobra-equipped 403.IAP, 872.ShAP with its Il-2s and 113.BAD, flying Il-4s.

Despite three solid weeks of action, HLeLv 24 could still muster 23 serviceable Messerschmitts on the morning of 1 July, and a further four brand new G-6s arrived at Lappeenranta that afternoon.

Taking full advantage of the light summer nights, the unit had tasked 3/HLeLv 24 with providing aerial cover for a convoy of troop ships sailing through the Gulf of Finland near Teikarsaari between 0400 and 0500 hrs on the 1st. Midway through the patrol ten Il-2s, escorted by thirteen La-5s, attempted to attack the Finnish vessels, but were swiftly driven away by four Bf 109s. The remaining Messerschmitt pilots engaged the Lavochkin fighters, shooting four of them down.

BASE RAIDS

Both HLeLv 24 at Lappeenranta and HLeLv 34 at Taipalsaari had escaped Soviet attention throughout June, as had the Luftwaffe units at Immola. However, at 1955 hrs on 2 July 35 Pe-2s and 40 Il-2s, escorted by 20 fighters, attacked Lappeenranta. Fortunately, radio intelligence

3/HLeLv 24's pilots pose at Lappeenranta for a flight photograph on 10 July 1944. They are, from left to right, 2Lt Per-Erik Ohls (2 kills), 1Lt Jorma Saarinen (23 kills), flight leader 1Lt Kyösti Karhila (32 kills), Sgt Kosti Koskinen (2 kills), SSgt Leo Ahokas (12 kills), SSgt Emil Vesa (29.5 kills), 2Lt Erkki Estama, MSgt Jouko Huotari (17.5 kills) and Sgt Risto Helava (4 kills). The latter pilot was shot down (in MT-440) over Heinjoki by a P-39 from 196.IAP just 24 hours after this group photo was taken. Helava was quickly captured by Russian troops (*SA-kuva*)

3/HLeLv 24's MSgt Nils Katajainen taxies across the grass at Lappeenranta in Bf 109G-8 Wk-Nr 200041 on 30 June 1944. Moments later, he took off on a sortie that would see him destroy an Il-4 bomber, and thus increase his overall tally to 30.5 kills. Flying one of only two G-8s supplied to the Finns, Katajainen claimed eight victories in just five days – 29 June to 3 July. On the latter date the *Gustav* was badly shot up by a Yak-9 just after its pilot had downed another Yakovlev fighter. Katajainen was left with little choice but to force-land at Nuijamaa. Supplied as an attrition replacement, this aircraft survived for only a short time in frontline service. Indeed, its service life was so brief that groundcrews at Lappeenranta didn't even have the chance to paint out the G-8's delivery markings – its Finnish Air Force serial, MT-462, was actually chalked on behind the fighter's cockpit! (*SA-kuva*)

alerted HLeLv 24 of the impending raid, allowing the unit to get 11 freshly refuelled and re-armed Bf 109Gs airborne just 30 minutes before the airfield was attacked. However, these aircraft were lured away from the main force by a decoy formation, thus allowing the Soviet bombers to drop their ordnance without interference.

A swarm of 2nd Flight fighters had just recovered back at Lappeenranta after completing a patrol when the enemy bombers struck. Caught out in the open, MT-246 and MT-450 were destroyed and four other fighters lightly damaged. Two captured Pe-2 photo-reconnaissance aircraft from PLeLv 48 that were stationed at the base were burnt out as well.

Eight of HLeLv 34's Bf 109s had also been scrambled in response to the raid, and they arrived over Lappeenranta within five minutes of taking off. Those pilots lured away by the decoy formation quickly realised their error, and they joined HLeLv 34 in attacking the ranks of vulnerable Il-2s that were strafing their base. The two units jointly claimed 11 Ilyushins from 448., 703. and 872.ShAP (all members of 281.ShAD) shot down. Chasing the now retreating Soviet formation eastward beyond Viipuri, the Finnish pilots destroyed a further four Pe-2s of 276.BAD and a solitary Yak-9.

Despite the ferocity of these attacks, neither the Finns at Lappeenranta or *Gefechtsverband Kuhlmey* at nearby Immola (bombed during the same raid) had had their ability to generate combat patrols seriously affected.

By early July the Russian offensive had stalled near Tali and Ihantala, so the Red Army chose instead to launch its final attack further east between Vuosalmi and Äyräpää. Fresh troops flooded across the River Vuoksi, and succeeded in establishing a small bridgehead on the west shore of the Gulf of Viipuri. This stronghold immediately became the target for a concentrated artillery and bombing campaign that lasted for a full two weeks, following which the Soviets retreated back across the Gulf of Viipuri and all offensive activity on the Karelian Isthmus ceased.

During the campaign to wipe out the communist bridgehead, HLeLv 24 had provided fighter escorts for Finnish bomber formations, and large scale combats became less frequent.

The Red Army launched its final assault on 3 July, and that very morning nine Bf 109Gs (led by 1Lt Joel Savonen) engaged 40 Il-4s, 40 Il-2s and 30 escorting fighters between Tali and Ihantala. On this occasion the Russians held their own, losing only one bomber and two fighters. The Finns, in return, had 19.5-kill ace WO Viktor Pyötsiä shot down when both he and his Bf 109G-2 (MT-235) were struck by defensive

On the morning of 19 June 1944 a Luftwaffe Savoia Marchetti S.81 of *Transportgruppe* 10 flew into Lappeenranta and picked up Capt Hans Wind and a number of other pilots from 3/HLeLv 24. From here it flew them to Insterburg, in Germany, where they collected a handful of brand new Bf 109G-6s. Returning immediately to Finland, the flight saw action with its new fighters that same evening, Capt Wind (in MT-439) scoring three victories. Here, war correspondent Erik Blomberg is helped aboard the S.81 by the 3rd Flight commander (*SA-kuva*)

Bf 109G-6/R6 *Kanonenboote* MT-465 'Yellow 7' was one of 14 *Gustavs* delivered with an additional pair of 20 mm cannon fitted in underwing gondolas. Assigned to 2/HLeLv 24, and seen here at Lappeenranta in July 1944, this aircraft was stripped of its bulky gun pods on the express orders of its regular pilot, 1Lt Atte Nyman. Most other *Kanonenboote* also had their bulky wing guns removed, as the Finnish pilots found the fighter's standard armament of one 20 mm cannon and two 13 mm machine guns more than adequate for the job if an attack was pressed home from close range. As if to prove this point, Nyman scored his fifth, and last, kill (an Il-2) in this very machine on 29 June – in 18 months of service with HLeLv 24, he had flown 150 sorties (*A Nyman*)

fire from an Il-2 of 277.ShAD. Succeeding in downing the Ilyushin, the Winter War veteran then took to his parachute over Finnish territory.

A short time later 1Lt 'Kössi' Karhila's flight was patrolling the same area when it met 'only' 15 *Stormoviks*, escorted by 10 La-5s. Enjoying better odds than Savonen's men had done earlier in the day, the Finnish pilots duly claimed three assault aircraft and four fighters shot down. Russian losses are known to have included Il-2s of 448. and 872.ShAP, as well as fighters from 14., 29., 159. and 191.IAP. The flight also lost a Bf 109, however, when MSgt 'Nipa' Katajainen's MT-462 was hit, causing the ace to belly land his fighter – Katajainen had claimed four kills during this engagement, taking his tally to 34.5.

The 3rd also saw Capt Aate Lassila and 1Lt Erik Teromaa formally given control of the 1st and 2nd Flights, respectively.

A spell of heavy rain on 4 July brought some respite for HLeLv 24, although a single interception and a bomber escort mission netted the unit two fighters and one ground attack aeroplane destroyed. The following day 1Lt Karhila led eight Bf 109s on a sweep of the frontline in the vicinity of Koivisto, where ten *Stormoviks* of 13.AAE, KBF (artillery fire-control unit), escorted by a dozen Yak-9s from 13.KIAP, were encountered.

One swarm attacked the assault aircraft, sending two of them down, while the remaining four pilots engaged the escorts and destroyed three fighters.

Once again HLeLv 24 failed to escape unscathed, however, for MSgt Katajainen's MT-476 was struck by a single cannon shell. The ace had just downed a Yak-9 when the round exploded against his head armour and almost knocked him out. Barely conscious, he instinctively headed for Lappeenranta, where he tried to belly land his machine at 500 km/h. The fighter hit the ground with a thump, bounced more than 200 metres, then struck the ground again. This time it travelled a much shorter distance and duly flipped over when its nose dug in. The engine broke loose and the rest of the machine ground to a steaming halt, the pilot having been thrown from the wreck – a rescue team found him bloodied and dusty in a state of unconsciousness, yet Nils Katajainen was still alive. His war was over.

From the 6 to 8 July HLeLv 24 performed a series of bomber escort missions, claiming six more kills during this period. Then, on the afternoon of the 9th, Capt Veikko Ala-Panula (on temporary duty from HLeLv 28) led seven Messerschmitts on a sweep of Äyräpää in support of troops on the ground. The swarm was bounced by eight Yak-9s from 14.GIAP over the target area, and after a short clash the Russians retreated minus three of their number. Several hours later, whilst performing a similar mission, 1Lt Väinö Suhonen's eight Bf 109s engaged a mixed gaggle of eleven fighters, claiming one La-5 and one Yak-9 shot down.

At midday on 10 July eight Messerschmitts again flew in support of soldiers fighting in the vicinity of Äyräpää. This time they were attacked by 15 fighters, and once more the Finns emerged victorious, claiming five La-5s and one Yak-9 shot down – 3rd Flight leader 1Lt Karhila (in MT-461) was credited with one of each. The final sorties of the day saw Capt Ala-Panula leading 12 fighters on an evening bomber escort mission to Äyräpää. As the bombers neared the target 20 Soviet fighters tried to effect an interception, but the Finnish pilots stood their ground and destroyed six La-5s. Three were claimed by 1Lt 'Olli' Puro (in MT-479), whose success prompted the 2nd Flight adjutant to write in the unit's diary, '1Lt Puro has returned to his flight with a leg still in plaster and a thirst for blood'. Both 159. and 191.IAP are known to have suffered casualties.

The second of just two photo-recce equipped Bf 109G-8s supplied to the Finns, MT-483 had its cameras removed and served exclusively as an interceptor with 1/HLeLv 24. Flown to Finland on 12 July 1944, the aircraft was assigned to 1st Flight Leader Capt Aate Lassila. Seen here at Lappeenranta shortly after flying in from Germany, MT-483 saw very little action due to the lateness of its arrival in Finland (*J Hyvönen*)

MSgt Nils 'Nipa' Katajainen checks his maps prior to starting up MT-462 at Lappeenranta on 30 June 1944. A 3rd Flight veteran who served with the unit for much of the Continuation War (he completed a seven-month tour on a maritime patrol unit in 1942-43), Katajainen's frontline career came to an end on 5 July 1944 when he crashed MT-476 at high speed after it had suffered serious combat damage. Badly injured in the forced landing, the 35.5-kill ace was hospitalised for the rest of the conflict. A veteran of 196 combat sorties, Katajainen became the only reservist fighter pilot to be awarded the Mannerheim Cross on 21 December 1944 (*SA-kuva*)

On the 11th 1Lt Karhila's flight twice escorted bombers to Äyräpää, with the sole kill (an La-5) of the morning sortie falling to the flight leader in MT-461. The noon mission saw a more concentrated effort by Soviet fighters to prevent the bombers from reaching their target, and although the escorts destroyed two Yak-9s, Sgt Risto Helava was forced to bail out of MT-440 after it received a series of hits from an Airacobra of 196.IAP. He was quickly captured.

No sorties were flown on 12 July, and during the 48-hour period which followed, two more escort missions were completed without incident. Protecting LeR 4's modest bomber force was a very important job, and it restricted pilots from chasing after Soviet fighters. Aware of their responsibilities, the usually free-spirited Messerschmitt pilots performed this mission with such discipline over the Karelian Isthmus that not a single bomber was downed by enemy fighters during this phase of the campaign.

The 15th proved to be the last day in which large-scale dogfights took place between similar numbers of opposing fighters. The two Messerschmitt squadrons claimed twelve Soviet aircraft destroyed during the course of five missions, the first of these involving Bf 109s from HLeLv 24 (2nd and 3rd Flights) taking place over Äyräpää. A flight of Il-2s, escorted by 20 fighters, was intercepted, and one *Stormovik* from 7.GShAP and four Yak-9s from 14.GIAP were downed.

Three days later 16 Bf 109s escorted 30 bombers sent to strike at Äyräpää once again. Nearing the town, six La-5s of 159.IAP made a diving attack on the bombers, but 1Lt Karhila's eight-strong swarm intervened and disposed of two of the Lavochkins. However, the 3rd Flight's deputy leader, 1Lt Jorma Saarinen, was hit during the engagement soon after the 23-kill ace had downed one of the La-5s. Deciding to belly land his aeroplane (MT-478) in a nearby field, Saarinen failed to spot a road that ran through the open ground and he crashed into a bank and was killed. He was the last pilot from HLeLv 24 to die in combat.

That afternoon the unit flew its final escort mission to Äyräpää, 16 fighters protecting 24 bombers that attacked the town without incident.

The Finnish Army's defensive line between Tali and Ihantala had held firm, despite being attacked by a numerically superior force. This meant that the Red Army's attempt to cross the Gulf of Viipuri could never succeed, and its final push westward between Vuosalmi and Äyräpää failed to materialise after Stalin called off the offensive on 12 July. The fighting nevertheless continued for a further six days until all communist troop reserves had been exhausted.

1Lt Kyösti 'Kössi' Karhila poses in the cockpit of MT-461 'Yellow 6' at Lappeenranta on 13 July 1944. He had commanded the 3rd Flight for just a fortnight when this shot was taken, and he was duly posted to command 2/HLeLv 30 on 21 July 1944. Karhila chose to keep the wing cannon in MT-461, which he used to score eight of the 32.5 kills he accrued whilst flying 304 sorties (*SA-kuva*)

The effective rearguard action fought by the Finns, combined with the successful Allied landings in Normandy early the previous month, greatly influenced Stalin in his decision to halt the invasion and focus instead on the push for Berlin.

During the 38 days of the 'Great Attack', the Messerschmitt pilots of both HLeLv 24 and 34 had claimed no fewer than 425 aircraft shot down and 78 damaged during 355 missions (2168 sorties). The Luftwaffe's II./JG 54 – part of *Gefechtsverband Kuhlmey* – also reported having destroyed a further 126 aircraft during the same period in the same operational area, the unit flying 179 missions (984 sorties). Finally, Finnish anti-aircraft batteries accounted for an additional 400 aircraft. The Soviet air forces had suffered truly stunning losses, and in spite of the continuous arrival of new aircraft, the once mighty 13th Air Army possessed just 800 aircraft at the conclusion of the fighting over the Karelian Isthmus.

Only ten Finnish Messerschmitts were lost to Soviet fighters, three posted missing in action, three more to flak and two to 'assault' aircraft. Eight pilots were killed and three captured.

TOWARDS PEACE

From 19 July onwards only small formations of Soviet aircraft ventured over the frontline. On this day, one such patrol was intercepted by elements of 2nd Flight, which destroyed four fighters. Twenty-four hours later, six Messerschmitts, led by 1Lt Erik Teromaa, reconnoitred the front from Vuosalmi to Tali. Overflying Äyräpää, they bounced five La-5s of 159.IAP, claiming three shot down for the loss of 1Lt Toimi Juvonen (on secondment from HLeLv 28) in MT-475. His engine had been hit by cannon fire during the dogfight, and it duly seized on the return flight to base. Attempting to force-land the Bf 109, Juvonen crashed to his death when his fighter came down in a forest.

Looking every inch the 'steely-eyed' fighter ace, SSgt Emil 'Emppu' Vesa sits strapped into the cockpit of MT-460 at Lappeenranta on 30 June 1944. He flew with the 3rd Flight from 3 December 1941 until war's end, completing 198 sorties and claiming 29.5 kills. He scored his last eight victories in MT-460 (*SA-kuva*)

1Lt Olavi 'Olli' Puro poses in the cockpit of MT-479 at Lappeenranta on 10 July 1944. He had just returned to his unit after suffering wounds in action on 28 June – one of his legs was still in plaster when this shot was taken. A member of the 2nd Flight from 4 April 1943, Puro scored 36 kills (four in MT-479) in 207 sorties. His tally would have undoubtedly been higher had he not missed almost two weeks of flying at the height of the offensive (*SA-kuva*)

On 22 July 1Lt Puro and his wingman engaged fifty Il-2s, nine Pe-2s and twenty escorting La-5s whilst on a reconnaissance mission to Seiskari. Despite being massively outnumbered, the Finns dived headlong into the unsuspecting Soviet formation, and Puro (in MT-461) swiftly downed two Il-2s and an La-5 into the Gulf of Finland. Boosting his score to 35 kills, these victories made Puro the squadron's most successful serving pilot. The following day he added another La-5 to his tally.

HLeLv 24's final combat in a long war was fought on 25 July, when the 3rd Flight's 1Lt Väinö Suhonen led a swarm of fighters against 30 Il-2s from 7.GShAP, escorted by 20 La-5s. Caught near Someri, heading east across the Gulf of Finland, three *Stormoviks* and two fighters were shot down – Suhonen claimed three of these in MT-461. This same formation was also attacked by a flight from HLeLv 34 (led by Maj Luukkanen), which destroyed two Il-2s and two fighters.

On 1 August HLeLv 24 had 18 serviceable Bf 109G-6 fighters on strength, its nine surviving G-2s having been handed over to LeR 2's HLeLv 28 towards the end of July.

Maj Jorma Karhunen, HLeLv 24's Commanding Officer, wrote the following summary of his unit's achievements during a hectic summer to his commanding officer, Col Gustaf Magnusson, on 31 August;

'When military activity increased on the Finnish front, the squadron was prepared, within its limits, to repel the superior air forces of the enemy.

'By the early spring, training flights with the Bf 109G-2 had been initiated, and come the end of May all of our Brewsters had been transferred to other units, leaving us equipped exclusively with Bf 109Gs. From a technical viewpoint, these new machines were far more difficult to overhaul than our previous

Sgt Arvo Koskelainen of 1/HLeLv 24 sits atop MT-506 'Yellow 8' at Utti in September 1944. He had used this late-delivery *Gustav* to down one of the five kills he claimed whilst flying 140 sorties with the unit. On 10 November 1944 Koskelainen, like all other reserve pilots in the air force, ended his combat tour and returned to civilian life (*O Leskinen*)

1/HLeLv 24's 2Lt Heimo 'Hemmi' Lampi (13.5 kills), 1Lt Mikko Pasila (10 kills) and 1Lt Otso Leskinen (1 kill) are seen with MT-477 'Yellow 7' just days after the Continuation War had come to an end following the signature of an armistice on 4 September 1944. Within a fortnight the squadron had been transferred to Utti. 'Hemmi' Lampi had claimed his final victory (an La-5) with MT-477 on 10 July (*O Leskinen*)

fighters, but the squadron's technical personnel succeeded in tackling any problems that arose. For this, they are to be commended. During the various engagements of May and early June, my pilots gained the necessary experience required to allow them to trust the new fighter type in combat.

'When the enemy's attacking force succeeded in flying around 1500 sorties per day (which it did on several days), it was natural that my pilots should begin to show signs of fatigue after having to fly virtually round the clock. Some of the losses suffered by the unit can be attributed to this cause. Despite being totally outnumbered, my pilots did an excellent job.

'The ferocity of the battle is grimly underlined by the number of casualties that we suffered. The loss of flight leaders Myllymäki and Nissinen and deputy leaders Sarjamo and Saarinen, plus the wounding of Capt Wind, 1Lt Puro and WO Pyötsiä, hurt us badly.

'The technical and support staff in the squadron have fulfilled all tasks requested of them, which became very laborious during base transfers.

'The good fighting efficiency of the unit depended on the competence of all personnel in every section, good comradeship and mutual trust.

'When I began my review the squadron personnel at Suulajärvi numbered 31 officers, 121 non-commissioned officers and 71 enlisted men. We also had 14 aircraft on strength.

'During the review period, the various flights within the unit completed the following number of sorties:

1st Flight 150 missions (436 sorties)
2nd Flight 123 missions (412 sorties)
3rd Flight 136 missions (424 sorties)

Whilst undertaking these missions, squadron pilots claimed 240 confirmed aerial victories, 33 aircraft shot down that could not be confirmed and 32 seriously damaged.

The squadron's losses were as follows – 8 officers and 1 NCO killed or missing in action, 2 officers and 4 NCOs wounded and 13 aircraft lost,

All seven of 1/HLeLv 24 *Gustavs* are seen lined up in the open at Utti in mid September 1944. Such a photograph could never have been taken earlier in the month, when the unit was still very much at war with the Soviet Union. The fighter closest to the camera is MT-457, which was issued to 1/HLeLv 24 from 1/HLeLv 34 on 1 August 1944. Whilst with the latter unit, it had been used by 'ace of aces' WO Ilmari Juutilainen to claim the last 18 of his 94 victories (*J Hyvönen*)

two of which were destroyed in an air raid. A further 17 aeroplanes were damaged.'

The mission shift from interceptions in June to bomber escorting in July is clearly revealed in the squadron's operational records, for in June 482 interception and 196 escort sorties were flown, whilst in July 320 interception and 403 escort sorties were completed.

ARMISTICE

Having repelled the communist invasion, Finland now sought a permanent peace with the Soviet Union, and this led to an armistice on 4 September 1944 and the signing of a truce in Moscow two weeks later. The terms of the latter document included the removal of German troops from northern Finland, resulting in a short land war being fought in Lapland between the Finnish Army and its former ally. In February 1947 the truce was ratified in the Paris Peace Treaty, forcing Finland to hand over the same chunks of land as had been surrendered in 1940, plus the town of Petsamo on the Arctic Sea coast. These demands were hardly fair, for Soviet troops had never even got close to seizing any of these areas. What had been won on the battlefield had been lost in the peace negotiations.

On 12 September 1944 HLeLv 24's 1st and 3rd Flights had transferred to Utti, followed by the 2nd Flight one week later. All 25 of its Bf 109G-6s were now sited at the same base that the unit had commenced fighting the Winter War from almost five years earlier. The squadron was grounded, and the yellow theatre markings on its fighters removed.

One of the conditions of the truce saw the armed forces demobilised on 4 December, with all reservists being sent back to civilian life. At the same

Retired Bf 109G-6s repose at Utti in late September 1944. In the foreground is MT-431, which had belonged to the 2nd Flight, while MT-441 'Yellow 1' was a 3rd Flight machine. According to the terms of the armistice, all yellow eastern front markings had to be removed by 14 September 1944 (*V Lakio*)

1/HLeLv 24's MT-504 'Yellow 1' sits on the grass at Utti in late September 1944. One of the last *Gustavs* flown in to Finland from Anklam, in Germany, the fighter had arrived at the air force's allocation depot on 25 August, and had been assigned to the squadron on armistice day. Despite the war being over, HLeLv 24 still adorned the Bf 109G with an individual tactical number forward of the cockpit just in case it was called to arms again (*V Lakio*)

time the air force's squadrons were re-numbered, and *Hävittäjä Lentolaivue* 24 duly became HLeLv 31. Still in existence today, the unit is presently equipped with F/A-18 Hornets as the flying element of the Karelian Air Command.

On 22 December 1944 the last Mannerheim Cross (number 170) to be awarded to a member of HLeLv 24 was presented to MSgt Nils Katajainen, who was also the only reservist fighter pilot to receive the medal. Both 1Lts Kyösti Karhila and Olavi Puro had achieved scores similar to Katajainen, but the quota of medal winners had already been filled.

Pilots from *Hävittäjä Lentolaivue* 24 had been credited with shooting down 883 aircraft during the two conflicts. In return, the unit had lost 55 fighters to all causes, 38 of which were shot down in combat. Twenty-nine pilots had been either killed or posted missing in action, and a further three were captured (these figures also include the claims and losses of other units controlled by HLeLv 24). Finally, the unit produced five direct and two indirect Mannerheim Cross winners, which was more than any other squadron in the Finnish Air Force – truly a sign of an elite unit.

LEADING Bf 109G ACES OF HLeLv 24

Rank	Name	Flight	Victories
Capt	Wind, Hans**	3	36
1Lt	Puro, Olavi	2, 3	28.5
MSgt	Vesa, Emil	3	20
1Lt	Saarinen, Jorma(+)	2, 3	18
MSgt	Katajainen, Nils*	3	18
SSgt	Järvi, Tapio	2	17
Sgt	Halonen, Eero	2	16
1Lt	Suhonen, Väinö	1	15
1Lt	Karhila, Kyösti	3	13
1Lt	Riihikallio, Eero	2	10

** - double Mannerheim Cross
* - Mannerheim Cross
+ - killed in action

APPENDICES

APPENDIX 1

COMMANDING OFFICERS

Squadron commanders

Capt Magnusson, Gustaf Erik (Maj 06/12/39, Lt Col 10/11/41)	21/11/38 - 31/05/43
Capt Karhunen, Jorma (Maj 31/8/43)	01/06/43 - 04/12/44

1st Flight wartime leaders

Capt Carlsson, Eino	30/11/39 - 13/03/40
Capt Luukkanen, Eino (Maj 01/11/42)	25/06/41 - 10/11/42
1Lt Wind, Hans	11/11/42 - 15/01/43
Capt Sarvanto, Jorma	16/01/43 - 07/07/43
1Lt Nissinen, Lauri++	08/07/43 - 17/06/44
1Lt Savonen, Joel	17/06/44 - 02/07/44
Capt Lassila, Aate	03/07/44 - 04/09/44

2nd Flight wartime leaders

1Lt Vuorela, Jaakko+	30/11/39 - 30/01/40
1Lt Karhunen, Jorma	01/02/40 - 13/03/40
Capt Ahola, Leo	25/06/41 - 23/06/42
Capt Ervi, Pauli	24/06/42 - 10/02/43
1Lt Törrönen, Iikka++ (Capt 13/4/43)	11/02/43 - 02/05/43
1Lt Lumme, Aulis	02/05/43 - 24/08/43
Capt Myllymäki, Jouko++	25/08/43 - 25/06/44
1Lt Lumme, Aulis	25/06/44 - 02/07/44
1Lt Teromaa, Erik	03/07/44 - 29/07/44
Capt Linkola, Mikko	30/07/44 - 04/09/44

3rd Flight wartime leaders

1Lt Luukkanen, Eino (Capt 15/02/40)	30/11/39 - 13/03/40
1Lt Karhunen, Jorma (Capt 4/8/41)	25/06/41 - 26/05/43
1Lt Wind, Hans (Capt 19/10/43)	27/05/43 - 28/06/44
1Lt Karhila, Kyösti	30/06/44 - 20/07/44
1Lt Suhonen, Väinö	21/07/44 - 04/09/44

4th Flight wartime leaders

Capt Magnusson, Gustaf (Maj 06/12/39)	30/11/39 - 13/03/40
1Lt Sovelius, Per-Erik (Capt 04/08/41)	25/06/41 - 14/02/42
1Lt Törrönen, Iikka	15/02/42 - 11/02/43 (disbanded)

5th Flight wartime leader

1Lt Ahola, Leo	30/11/39 - 13/03/40 (disbanded)

Notes

\+ killed in flying accident
++ killed in action

APPENDIX 2

OPERATIONAL LOSSES

Date	Aeroplane	Pilot	Circumstances
23/12/39	FR-111	Sgt Tauno Kaarma, wounded	damaged by 25.IAP I-16, crashed at Lyykylänjärvi
20/01/40	FR-107	SSgt Pentti Tilli, killed	49.IAP I-16, shot down over Sääksjärvi
01/02/40	FR-115	2Lt Tapani Harmaja, killed	7.IAP I-16, shot down over Venäjänsaari
02/02/40	FR-81	1Lt Fritz Rasmussen, killed	25.IAP I-16, shot down over Rauha
10/02/40	FR-102	MSgt Väinö Ikonen, wounded	damaged by 7. or 25.IAP I-16, crashed at Simola
19/02/40	FR-80	1Lt Erhard Frijs, killed	25.IAP I-16, shot down over Heinjoki
26/02/40	FR-85	Sgt Tauno Kaarma, no injuries	damaged by 68.OIAP I-16, crashed at Immola
29/02/40	FR-94	1Lt Tatu Huhanantti, killed	68.OIAP I-16, shot down over Ruokolahti
05/03/40	FR-76	Sgt Mauno Fräntilä, wounded	damaged by 7.IAP I-16, crashed at Virolahti
03/12/41	BW-385	1Lt Henrik Elfving, killed	AA, shot down over Novinka
24/01/42	BW-358	Sgt Eino Myllymäki, killed	152.IAP Hurricane, shot down over Belomorsk
26/02/42	BW-359	Sgt Tauno Heinonen, bailed out	MiG-3 (unit unknown), shot down over Liistepohja
09/03/42	BW-362	Sgt Paavo Mellin, captured	152.IAP Hurricane, shot down over Uikujärvi
29/05/42	BW-390	–	air raid at Nurmoila
08/06/42	BW-394	1Lt Uolevi Alvesalo, no injuries	damaged by 152.IAP Hurricane, crashed at Rukajärvi
25/06/42	BW-372	1Lt Lauri Pekuri, injured	damaged by 609.IAP Hurricane, ditched off Seesjärvi
25/06/42	BW-381	Sgt Kalevi Anttila, injured	damaged by 609.IAP Hurricane, crashed at Seesjärvi
18/08/42	BW-378	2Lt Aarno Raitio, killed	71.IAP, KBF I-16, shot down over Kronstadt
30/10/42	BW-376	Sgt Paavo Tolonen, killed	71.IAP, KBF I-16, shot down over Oranienbaum
21/04/43	BW-354	SSgt Tauno Heinonen, killed	4.GIAP, KBF La-5, shot down over Oranienbaum
21/04/43	BW-352	WO Eero Kinnunen, killed	AA, shot down over Oranienbaum
02/05/43	BW-380	Capt Iikka Törrönen, killed	3.GIAP, KBF LaGG-3, shot down over Oranienbaum
04/05/43	BW-388	Sgt Jouko Lilja, killed	3.GIAP, KBF LaGG-3, shot down over Seivästö
17/06/43	BW-351	–	air raid at Suulajärvi
31/08/43	BW-356	Sgt Sulo Lehtiö, killed	13.KIAP, KBF Yak-7B, shot down over Koivisto
10/11/43	BW-366	1Lt Vilppu Perkko, captured	13.KIAP, KBF Yak-7, shot down over Oranienbaum
02/06/44	MT-204	1Lt Heikki Herrala, killed	14. or 29.GIAP Yak-9, shot down over Kivennapa
07/06/44	MT-225	SSgt Viljo Kauppinen, wounded	collided with 196.IAP P-39, crashed at Pilppula
17/06/44	MT-227	1Lt Urho Sarjamo, killed	159.IAP La-5, shot down over Perkjärvi
17/06/44	MT-229	1Lt Lauri Nissinen, killed	hit by Bf 109G-2 MT-227

Date	Aeroplane	Pilot	Circumstances
22/06/44	MT-442	2Lt Erkki Nukarinen, killed	14. or 29.GIAP Yak-9, shot down over Tali
29/06/44	MT-439	1Lt Ahti Laitinen, captured	159.IAP La-5, shot down over Ihantala
02/07/44	MT-246	–	air raid at Lappeenranta
02/07/44	MT-450	–	air raid at Lappeenranta
03/07/44	MT-235	WO Viktor Pyötsiä, wounded	277.ShAD Il-2, shot down over Nuijamaa
05/07/44	MT-476	MSgt Nils Katajainen, wounded	damaged by 13.KIAP, KBF Yak-9, crashed at Lappeenranta
11/07/44	MT-440	Sgt Risto Helava, captured	196.IAP P-39, shot down over Heinjoki
18/07/44	MT-478	1Lt Jorma Saarinen, killed	damaged by 159.IAP La-5, crashed at Antrea
20/07/44	MT-475	1Lt Toimi Juvonen, killed	damaged by 159.IAP La-5, crashed at Joutseno

APPENDIX 3

SQUADRON ACES

Rank	Name	LeLv 24 Kills	FR	BW	MT	Total overall	Remarks
Capt	Wind, Hans**	75	–	39	36	75	Wounded 28/06/44
WO	Juutilainen, Ilmari*	36	2	34	–	94	Second MHR in HLeLv 34
MSgt	Katajainen, Nils*	35.5	–	17.5	18	35.5	Wounded 05/07/44
1Lt	Puro, Olavi	34	–	5.5	28.5	36	
1Lt	Nissinen Lauri*	32.5	4	22.5	6	32.5	KIA 17/06/44
Capt	Karhunen, Jorma*	31	4.5	26.5	–	31	
MSgt	Vesa, Emil	29.5	–	9.5	20	29.5	
SSgt	Järvi, Tapio	28.5	–	11.5	17	28.5	
1Lt	Saarinen, Jorma	23	–	5	18	23	KIA 18/07/44
WO	Kinnunen, Eero	22.5	3.5	19	–	22.5	KIA 21/04/43
1Lt	Suhonen, Väinö	19.5	–	4.5	15	19.5	
WO	Pyötsiä, Viktor	19.5	7	8.5	3.5	19.5	Wounded 03/07/44
MSgt	Huotari, Jouko	17.5	–	9.5	8	17.5	
Capt	Luukkanen, Eino	17	–	2.5	14.5	56	MHR in HLeLv 34
Capt	Sarvanto, Jorma	17	13	4	–	17	
1Lt	Lumme, Aulis	16.5	–	11.5	5	16.5	
1Lt	Riihikallio, Eero	16.5	–	6.5	10	16.5	
Sgt	Halonen, Eero	16	–	–	16	16	
MSgt	Alho, Martti	15	1.5	13.5	–	15	KIFA 05/06/43
1Lt	Teromaa, Erik	14	–	8	6	19	
WO	Turkka, Yrjö	14	4.5	9.5	–	17	
1Lt	Kokko, Pekka	13.5	3.5	10	–	13.5	
2Lt	Lampi, Heimo	13.5	–	5.5	8	13.5	
1Lt	Karhila, Kyösti	13	–	–	13	32	
Capt	Sovelius, Per-Erik	12.5	5.5	7	–	12.5	
1Lt	Pekuri, Lauri	12.5	–	12.5	–	18.5	
1Lt	Sarjamo, Urho	12.5	–	6.5	6	12.5	KIA 18/07/44

Rank	Name	LeLv 24 Kills	FR	BW	MT	Total overall	Remarks
SSgt	Ahokas, Leo	12	–	7	5	12	
Capt	Törrönen, Iikka	11	0.5	10.5	–	11	
1Lt	Metsola, Kai	10.5	–	6.5	4	10.5	
1Lt	Laitinen, Ahti	10	–	2	8	10	PoW 29/06/44
1Lt	Pasila, Mikko	10	–	5	5	10	
SSgt	Kauppinen, Viljo	9.5	–	8.5	1	9.5	Wounded 07/06/44
1Lt	Savonen, Joel	8	–	7	1	8	
SSgt	Peltola, Eino	7.5	–	7.5	–	10.5	
1Lt	Huhanantti, Tatu	6	6	–	–	6	KIA 29/02/40
SSgt	Virta, Kelpo	6	6	–	–	6	KIFA 28/01/41
Sgt	Avikainen, Onni	6	–	6	–	6	
MSgt	Ikonen, Väinö	5.75	1.75	4	–	5.75	
Maj	Magnusson, Gustaf	5.5	4	1.5	–	5.5	MHR in LeR 3
1Lt	Kauppinen, Osmo	5.5	–	5.5	–	5.5	
WO	Rimminen, Veikko	5.5	1.5	5	–	5.5	
Sgt	Keskinummi, Kosti	5.5	–	0.5	5	5.5	
Sgt	Mellin, Paavo	5.5	–	5.5	–	5.5	PoW 09/03/42
1Lt	Lakio, Vilppu	5	–	5	–	5	
1Lt	Lindberg, Kim	5	–	5	–	5	
1Lt	Nyman, Atte	5	–	–	5	5	
1Lt	Nieminen, Urho #	5	5	–	–	11	
SSgt	Tilli, Pentti #	5	5	–	–	5	KIA 20/01/40
Sgt	Koskelainen, Arvo	5	–	–	5	5	

Notes

** double Mannerheim Cross winner
* Mannerheim Cross (MHR)
# member of LLv 26
KIA Killed In Action
KIFA Killed In Flying Accident
PoW Prisoner of War

COLOUR PLATES

1

Fokker D.XXI (c/n III/17) FR-110 'Blue 7' of WO Viktor Pyötsiä, 3/LLv 24, Joroinen, April 1940

Serving with 3/LLv 24, Pyötsiä scored 7.5 kills in FR-110 during the Winter War, his tally including two 'doubles' on 27 December 1939 and 20 January 1940. This fighter is one of only two known examples to have carried victory symbols during the Winter War – whether these kill markings were also applied to the port side of the fin remains unconfirmed. 'Isä-Vikki' ('Father-Vikki') was one of the 'old hands' of LLv 24, remaining with the unit throughout the five years of conflict. Aside from his success with the D.XXI, Pÿotsiä went on to claim kills with the Brewster Model 239 and the Bf 109G whilst serving with the 1st Flight during the Continuation War. He was credited with 19.5 kills in 437 sorties, the latter figure being bettered by just one other fighter pilot – 'Eikka' Luukkanen who flew 441 sorties.

2

Fokker D.XXI (c/n III/1) FR-97 'White 2' of 1Lt Jorma Sarvanto, 4/LLv 24, Utti, January 1940

Sarvanto became Finland's first ace when he used this aircraft to down six DB-3Ms of 6.DBAP during a four-minute action fought south of Utti on 6 January 1940. His exploits received much coverage in the global press, and 'Zamba' Sarvanto went on to become the leading ace of the Winter War. On 1 February 1940 he was made deputy leader of 1/LLv 24, and after claiming a further two victories in both FR-80 and FR-100, he had raised his score to 13 confirmed aerial victories in just three weeks. Sarvanto was then posted to Sweden to evaluate the new Brewster Model 239s that has just been reassembled for the air force, and he was still fulfilling this assignment when the Winter War came to an end.

3

Fokker D.XXI (c/n III/13) FR-112 'Black 7' of 1Lt Jorma Karhunen, 1/LLv 24, Immola, December 1939

'Joppe' Karhunen flew FR-112 for five weeks whilst serving as deputy leader of 1/LLv 24, scoring three and two shared kills during this time. His scoring run in the fighter came to an end on 3 January 1940 when FR-112 was damaged in a taxying accident with another D.XXI at Värtsilä, causing it to be sent away to the State Aircraft Factory at Tampere for repairs – his final score with the Fokker fighter was 4.5. On 30 January Karhunen was appointed commander of 2/LLv 24, althought he spent the rest of the Winter War test flying Brewsters in Sweden.

4

Fokker D.XXI (c/n III/3) FR-99 'Black 1' of Maj Gustaf Magnusson, CO of LLv 24, Joutseno, January 1940

'Eka' Magnusson assumed command of LLv 24 a year before the Winter War broke out. A firm believer in the 'finger-four' fighter formation, he imparted his beliefs with determination to all his pilots, and this duly led to his squadron having clear tactical superiority over its Soviet counterparts. Such an advantage in turn meant that a great number of aerial victories could be scored by a modest fighter force. Proving this point, Magnusson downed four bombers during the Winter War, and achieved 'acedom' whilst still in command of LLv 24 on 8 July 1941 when he claimed a DB-3M in Brewster BW-380. From late May 1943 he headed LeR 3, creating an indigenious early warning and fighter control system (without radar) which played a key role in repelling the Soviet summer offensive of 1944. On 26 June 1944 he won the Mannerheim Cross for these achievements.

5

Brewster Model 239 BW-390 'White 0' of 2Lt Kai Metsola, 1/LLv 24, Nurmoila, October 1941

'Kaius' Metsola served in 1/LLv 24 throughout the Continuation War, being assigned BW-390 early on in the campaign – he eventually claimed three kills with it. On 2 February 1942 Metsola was promoted to first lieutenant, which was the highest rank a reserve (non-cadet graduate) pilot could reach. On 29 May 1942 this machine was burnt to the ground during an air raid on Nurmoila. Prior to moving on to Messerschmitts, Metsola had scored 6.5 kills in the Brewster, the last of these in BW-367 on 9 November 1943. Note the fighter's tactical number '0', which denotes that it was a reinforcement airframe brought in to boost numbers, or as an attrition replacement. Most flights typically had eight aircraft on strength at any time, and these were numbered '1' to '8'.

6

Brewster Model 239 BW-357 'White 3' of 1Lt Jorma Sarvanto, 2/LLv 24, Rantasalmi, July 1941

Hero of the Winter War, 'Zamba' Sarvanto served as deputy leader of 2/LLv 24 for the first four months of the Continuation War, flying BW-357, and adding two victories to his earlier score of 13. Eventually promoted to captain, he took a staff position at air force headquarters and subsequently became the assistant to the Finnish military attaché in Germany. Returning to frontline service as the commander of LeLv 24's 1st Flight on 16 January 1943, Sarvanto claimed a brace of victories in BW-373 to bring his final tally to 17 kills from 251 sorties. On 9 July 1943 he was made head of the air war department of the air force's cadet school, where he stayed for the rest of the conflict.

7

Brewster Model 239 BW-368 'Orange 1' of SSgt Nils Katajainen, 3/LLv 24, Kontupohja, March 1942

Joining 3/LLv 24 on the eve of the outbreak of the Continuation War, Nils Katajainen remained with the unit until posted away to learn to fly twin-engined maritime patrol aircraft on 9 September 1942. By then he had been credited with 13 kills, seven of which had been claimed with BW-368. After a six-month tour flying anti-submarine missions in captured Tupolev SB bombers, Katajainen's application for a transfer back to fighters was approved on 9 April 1943, and he rejoined 3/LeLv 24. Upon his return, the ace was reassigned BW-368, and he went on to add a further 4.5 kills (1.5 in this aircraft) in Brewsters.

8

Brewster Model 239 BW-378 'Black 5' of Capt Per-Erik Sovelius, CO of 4/LLv 24, Lunkula, October 1941

Deputy leader of 4/LLv 24 during the Winter War, 'Pelle' Sovelius flew D.XXI FR-92 throughout the conflict, scoring 5.75 victories. Promoted to flight leader come the Continuation War, Sovelius's assigned mount was Model 239 BW-378 (a donation-funded machine, hence the titling below its cockpit). He scored his remaining seven kills in this fighter, and had completed 257 sorties by the time he accepted a staff position at air force headquarters on 16 February 1942. A little over five months later, on 30 July, Sovelius was made CO of the air force's Test Flight, where he remained until given command of HLeLv 28 on 30 May 1944. By now a major, he failed to add to his tally of 12.75 kills whilst flying MS.406s and then Bf 109G-2s during the last months of the war.

9

Brewster Model 239 BW-371 'White 1' of WO Viktor Pyötsiä, 1/LeLv 24, Suulajärvi, March 1943

As previously mentioned, 'Isä-Vikki' Pyötsiä served with 1/LeLv 24 throughout the Continuation War. He was assigned BW-371 in February 1943, and flew it until SSgt Kalevi Anttila crashed the fighter in bad weather on 22 February 1944 – the latter pilot was killed in the accident. Although not claiming any victories with this particular Model 239, Pyötsiä scored 7.5 kills in other Brewsters between June 1941 and March 1944. Following his unit's conversion to the Bf 109G-2 in April 1944, he was assigned MT-244. Pyötsiä had claimed 4.5 kills in *Gustavs* when, on 3 July, he was shot down by the return fire from an Il-2 whilst at the controls of MT-235. He bailed out successfully, but hit his head upon landing and was hospitalised for the rest of the war.

10

Brewster Model 239 BW-354 'White 6' of SSgt Heimo Lampi, 2/LeLv 24, Tiiksjärvi, September 1942

On 25 June 1941, during his very first combat mission, 'Hemmi' Lampi (then just a lowly corporal) shared in the destruction of five SB bombers with MSgt Kinnunen whilst flying this very machine. Remaining with 2/LeLv 24 until 8 January 1943, Lampi was then posted away to complete officer training. He returned to the unit on 15 June 1943, being sent to 1/LeLv 24, where, three months later, he was promoted to 2nd lieutenant. Lampi had the unique distinction of scoring *Lentolaivue* 24's first – and also the first kill of the Continuation War – and last Brewster victories (the latter took the form of an La-5 downed on 2 April 1944 in BW-382). Once he had completed his transition onto the Bf 109G, Lampi was assigned MT-232, although he claimed four of his eight Messerschmitt kills with MT-235. Successfully completing 268 sorties by war's end, 'Hemmi' Lampi claimed 13.5 aircraft destroyed.

11

Brewster Model 239 BW-393 'Orange 9' of Capt Hans Wind, CO of 3/HLeLv 24, Suulajärvi, April 1944

When Hans Wind left the 1st Flight on 27 May 1943 to take command of 3/LeLv 24, he took his favourite Brewster, BW-393, with him. Soon after the swap had taken place, the fighter's individual number was changed from 'White 7' (see profile 13) to 'Orange 9' as per the flight's assigned tactical markings (see caption for profile 13). Of Wind's 39 Brewster victories, 26 were claimed in BW-393, which he flew from December 1941 through to April 1944, when his squadron converted to the Bf 109G. MT-201 and MT-439 proved to be the most successful of his new mounts, and he added a further 36 kills (25 in ten days) to take his tally to 75 in 302 sorties.

12

Brewster Model 239 BW-370 'Black 4' of 1Lt Aulis Lumme, 4/LeLv 24, Römpötti, October 1942

Joining LLv 24 at the beginning of the Continuation War, Lumme was one of the few reserve pilots appointed to lead a flight – he commanded 2/LeLv 24 twice following the loss of its regular CO, remaining in charge until a replacement officer was found. Lumme flew BW-370 for two years, firstly with 4/LLv 24 and then from 11 February 1943 with 2/LeLv 24, during which time he claimed 4.5 kills with it. The machine later received the squadron's lynx emblem on its forward fuselage. By the end of hostilities Lumme had flown 287 sorties and scored 16.5 victories.

13

Brewster Model 239 BW-393 'White 7' of 1Lt Hans Wind, CO of 1/LeLv 24, Suulajärvi, January 1943

Twenty-two-year-old 1Lt 'Hasse' Wind joined 4/LLv 24 on 1 August 1941, and it took him eight months to become an ace – his fifth kill took the form of a shared victory on 29 March 1942 in BW-378. The following August he was posted to 1/LeLv 24, and his score quickly started to escalate whilst flying BW-393. On 10 November 1942 Wind took command of the flight, and by the time he became CO of 3/LeLv 24 on 27 May 1943, the fin of BW-393 showed 29 kills (see commentary for profile 11).

14

Brewster Model 239 BW-352 'White 2' of MSgt Eero Kinnunen, 2/LeLv 24, Tiiksjärvi, September 1942

'Lekkeri' Kinnunen scored 3.5 kills in the Winter War in D.XXI FR-109. Flying Brewster BW-352 by the outbreak of the Continuation War, he used this fighter during the first encounter of the conflict on 25 June 1941. This action saw five SB bombers destroyed, Kinnunen sharing the kills on this occasion with Cpl Heimo Lampi (see profile 10), and then going on to down a further two victories on his very next mission. He flew BW-352 throughout his career, firstly with 2/LLv 24 and, from 11 February 1943, 3/LeLv 24. On 21 April 1943 WO Kinnunen was downed by anti-aircraft fire at Oranienbaum, crashing to his death in this machine. He had scored 22.5 kills.

15

Brewster Model 239 BW-384 'Orange 3' of 2Lt Lauri Nissinen, 2/LeLv 24, Tiiksjärvi, May 1942

MSgt 'Lapra' Nissinen, who had claimed four kills in the Winter War flying D.XXI FR-98, saw his first action of the Continuation War with 3/LLv 24. Achieving ace status in BW-353 on 8 July 1941, he was subsequently assigned BW-384 on 12 August 1941. On 28 January 1942 Nissinen was transferred to 2/LLv 24 and sent to Viena to repel an ever increasing number of lend-lease Hurricanes flying from bases in Murmansk. On 8 June 1942 he scored his 20th kill of the Continuation War in BW-384, this tally including six Hurricanes. Nissinen then chose to become a regular officer, and he started his course at the cadet school on 1 July 1942. He was decorated with the Mannerheim Cross four days later (see commentary for profile 29).

16

Brewster Model 239 BW-377 'Black 1' of SSgt Tapio Järvi, 4/LeLv 24, Römpötti, October 1942

'Tappi' ('Shorty') Järvi was a member of LLv 24 from 11 August 1941 until war's end. On 11 February 1943 his flight was absorbed into 2/LeLv 24 due to a shortage of Brewsters, and it was with this flight that he scored most of his kills – Järvi used BW-377 to claim 7.5 aircraft destroyed. During the 1944 offensive, he flew mostly wing-cannon armed Bf 109G-6/R6 MT-450, downing ten Il-2s with it. By the time he was promoted to master sergeant on 16 July 1944, Järvi's scoring run had come to an end, the ace finishing with 28.5 kills from 247 sorties.

17

Brewster Model 239 BW-393 'White 7' of Maj Eino Luukkanen, CO of 1/LeLv 24, Römpötti, November 1942

'Eikka' Luukkanen commanded the 1st Flight from the start of the Continuation War, scoring 4.5 kills in his assigned BW-375. On 1 June 1942 he switched to BW-393 (which later became Hans Wind's mount) and duly claimed a further seven victories with it over the Gulf of Finland. Promoted to major on 7 November 1942, and with his tally standing at 17, Luukkanen was posted as CO to maritime reconnaissance squadron LeLv 30, which was also stationed at Römpötti. He was placed in charge of the air force's first Bf 109G unit – LeLv 34 – on 29 March 1943, and he remained its commanding officer until war's end. Flying a record 441 sorties, Luukkanen claimed 56 aircraft shot down, and won the Mannerheim Cross on 18 June 1944.

18

Brewster Model 239 BW-372 'White 5' of 1Lt Lauri Pekuri, 2/LeLv 24, Tiiksjärvi, June 1942

Posted to 2/LLv 24 on 3 September 1941, Lauri Pekuri was assigned BW-351 upon his arrival. When his flight was transferred to Tiiksjärvi in January 1942, he was promoted to the position of deputy leader and switched to BW-372. On 25 June 1942 'Lasse' Pekuri claimed two Hurricanes in this machine, but was in turn shot up by a third Hawker fighter and forced to ditch his burning Brewster into a small Karelian lake. Struggling to escape from his sinking fighter, Pekuri almost drowned from the weight of his sodden flight gear. He managed to reach dry land and then evaded through the wilderness back to his own lines. These Hurricanes were the last of 12.5 victories that he scored with the Model 239. On 9 February 1943 Pekuri was transferred to LeLv 34, and he eventually became leader of the unit's 1st Flight. On 16 June 1944 he was shot down in MT-420 and captured, returning to Finland on 25 December 1944. Pekuri flew 314 sorties and claimed 18.5 aerial victories. Returning to BW-372, on 6 August 1998 Pekuri's fighter was raised from its watery grave by a American-financed salvage team, the Brewster being in such good condition that it may eventually be restored to airworthiness. Presently in storage in Dublin, having been spirited out of Russia in controversial circumstances in December 1998, BW-372 (the world's sole original Brewster fighter) is presently being offered for sale by a Canadian company.

19

Brewster Model 239 BW-366 'Orange 6' of Capt Jorma Karhunen, CO of 3/LeLv 24, Suulajärvi, May 1943

BW-366 was assigned to Jorma Karhunen from the start of the Continuation War on 25 June 1941, and he claimed his 31st, and last kill, in this machine on 4 May 1943 – an I-153 over the Gulf of Finland. He led 3/LeLv 24 until 27 May 1943, when he was given command of the whole squadron. By then a Mannerheim Cross winner (awarded on 8 September 1942), 'Joppe' Karhunen remained the unit's CO until war's end.

20

Brewster Model 239 BW-386 'Black 3' of MSgt Sakari Ikonen, 4/LLv 24, Kontupohja, April 1942

A veteran of the Winter War (during which he had claimed 1.75 kills in D.XXI FR-102 up until he was shot down and wounded on 9 February 1940), Sakari Ikonen was still serving with the 4th Flight when the Continuation War erupted. He scored three kills in BW-386, and usually flew as a wingman to flight leader Capt Sovelius. On 1 February 1943 Ikonen was posted to the Air Fighting School as an instructor, where he remained for the rest of the conflict. He completed 204 sorties and scored 5.75 victories.

21

Bf 109G-2 (Wk-Nr 14784) MT-216 'Red 6' of 1Lt Mikko Pasila, 1/HLeLv 24, Suulajärvi, April 1944

Mikko Pasila joined 1/LLv 24 on 17 December 1941 following an initial spell with LLv 30. He would remain with the fighter unit for the rest of the Continuation War, scoring his first victory, in BW-382, on 13 October 1942 – by early May 1943 he had achieved ace status with five kills (all in Brewsters). In April 1944 Pasila's flight became the first within HLeLv 24 to replace its Brewsters with Bf 109G-2s, and he was duly assigned MT-216. His association with this aircraft did not last long, however, for on 18 May he was forced to carry out an emergency wheels-up landing when its engine quit in flight. The aircraft features the tactical number red '6' on its nose, this marking having been applied whilst the fighter was serving with its previous unit, HLeLv 34. Pasila went on to claim five kills with the Bf 109G, thus increasing his final tally to ten victories from 200 sorties (see the commentary for profile 34 for further details).

22

Bf 109G-2 (Wk-Nr 13393) MT-229 'Yellow 9' of 1Lt Väinö Suhonen, 1/HLeLv 24, Suulajärvi, April 1944

'Väiski' Suhonen served with 1/LLv 24 from 5 July 1941 until war's end. By the time his flight converted to the Bf 109G, he had scored 4.5 victories with Brewsters, and he 'made ace' in MT-229 on 30 May 1944 (he scored one other kill in this fighter six days later). Also adorned with an old HLeLv 34 tactical number, MT-229 was subsequently assigned to flight leader 1Lt Lauri Nissinen, who also claimed two kills with it prior to his death on 17 June. As previously detailed in this volume, MT-229 was hit by the remains of MT-227, and both Nissinen and fellow ace 1Lt Urho Sarjamo were killed. Following MT-229's reassignment to his flight leader, 'Väiski' Suhonen had initially flown MT-238 and then G-6 MT-461. On 21 July 1944 he became the acting flight leader of 3/HLeLv 24, and by the end of the conflict Suhonen had completed 261 sorties, and been credited with 19.5 victories (15 of which were achieved during the Soviet summer offensive of 1944).

23

Bf 109G-2 (Wk-Nr 10522) MT-221 of 1Lt Jorma Saarinen, 2/HLeLv 24, Suulajärvi, May 1944

Upon completing his flying training, 2Lt Jorma Saarinen was posted to LeLv 24 on 28 May 1942. Joining the 2nd Flight, he claimed his first kill on 16 April 1943 when he downed a LaGG-3 whilst flying BW-380. By October Saarinen had increased his tally to five with the Brewster fighter. Following conversion to the Bf 109G in May 1944, he was assigned MT-221, and he used it to claim three kills up to 25 June. The Messerschmitt was lost on this day when flight leader Capt Jouko Myllymäki crashed in poor weather and was killed. Saarinen was then assigned G-6 MT-452, and on 10 July he transferred to 3/HLeLv 24. On 18 July Jorma Saarinen became his unit's final operational casualty when he struck a road bank whilst trying to crash-land battle-damaged MT-478 into a field. He had claimed 23 aircraft destroyed in just 139 sorties prior to his death.

24

Bf 109G-2 (Wk-Nr 14754) MT-213 'White 3' of 1Lt Eero Riihikallio, 2/HLeLv 24, Suulajärvi, May 1944

Eero Riihikallio was posted to 2/LeLv 24 just prior to Christmas 1941, and it took him 11 months to score his first kill – a Tomahawk on 23 November 1942, flying BW-377. Within six months his score had risen to 6.5, however. During the Soviet summer offensive of 1944 he flew MT-213, which is seen here adorned with the new style of tactical numbering introduced on 22 May. Flying a 'mere' 110 sorties, Riihikallio nevertheless claimed 16.5 victories (three in MT-213).

25

Bf 109G-2 (Wk-Nr 10322) MT-231 'Yellow 1' of 1Lt Kai Metsola, 1/HLeLv 24, Lappeenranta, June 1944

On 8 April 1944 this fighter was issued to 1/HLeLv 24's 1Lt 'Kaius' Metsola, whose biographical details are given in the commentary for profile 5. The machine also features the new style of tactical numbering, where the position of the numeral denoted the flight to which the fighter belonged. MT-231 was the last G-2 to serve with the unit, being handed over to HLeLv 28 on 24 July 1944. Metsola flew all his missions with the 1st Flight, completing 296 sorties and scoring 10.5 kills (four with the Bf 109G).

26

Bf 109G-6 (Wk-Nr 164929) MT-441 'Yellow 1' of 1Lt Ahti Laitinen, 3/HLeLv 24, Lappeenranta, July 1944

Laitinen was posted to 3/LeLv 24 from training on 14 April 1943, and he scored his first kill on 31 August in BW-393. A second Brewster kill followed prior to his conversion onto the Bf 109G in April 1944. On 19 June Laitinen was assigned MT-441, and he used it to score six victories. On 29 June he was shot down in combat in MT-439, parachuting into captivity with serious wounds. As with all other PoWs, Laitinen was released on 25 December 1944. Although he flew just 75 sorties, he had claimed ten kills.

27

Bf 109G-6 (Wk-Nr 164982) MT-456 'Yellow 6' of 2Lt Otso Leskinen, 1/HLeLv 24, Lappeenranta, June 1944

MT-456 was issued to 1/HLeLv 24 on 25 June 1944, and it became the regular mount of 2Lt Leskinen. Four days later he scored his solitary aerial victory with it when he downed a Yak-9. Promoted thereafter to 1st lieutenant, Leskinen led his 'finger-four' on numerous bomber escort missions until the Soviet assault on the Karelian Isthmus finally came to an end on 18 July 1944.

28

Bf 109G-6 (Wk-Nr 165461) MT-476 'Yellow 7' of MSgt Nils Katajainen, 3/HLeLv 24, Lappeenranta, July 1944

By the time Nils Katajainen had converted to the Bf 109G in May 1944, his tally had risen to 17.5 kills (all scored with Brewsters as a member of 3/LeLv 24). Between 23 June and 5 July he added a further 18 kills to his total, predominantly in MT-436 or MT-462. On 3 July he forced landed the latter fighter at Nuijamaa after it was damaged in combat. The ace emerged unscathed, although he was not so lucky 48 hours later. Soon after downing a Yak-9, he was shot up in MT-476 by a second Yakovlev fighter and crashed at high speed upon reaching Lappeenranta. Having completed 196 sorties, Nils Katajainen claimed 35.5 kills, and was duly awarded the Mannerheim Cross.

29

Bf 109G-2 (Wk-Nr 13577) MT-225 'Yellow 5' of 1Lt Lauri Nissinen, CO of 1/HLeLv 24, Suulajärvi, May 1944

The first Bf 109 allocated to HLeLv 24, this aircraft was issued to Lauri Nissinen on 4 April 1944. Ten days later he used it to shoot down a German Ju 188 reconnaissance aircraft which had entered the Finnish airspace without prior notice, and without recognisable national markings. MT-225 was written off by 9.5 victory ace SSgt Viljo Kauppinen when he crash-landed following combat on 7 June. As previously mentioned, Lauri Nissinen was killed in the mid-air collision with the wreckage of 12.5 kill ace 1Lt Urho Sarjamo's MT-227 (minus its right wing, which had been shot off) on 17 June. He was the only Mannerheim Cross winner killed in action, and his score at the time of his death stood at 32.5 aircraft destroyed.

30

Bf 109G-6/R6 (Wk-Nr 165342) MT-461 'Yellow 6' of 1Lt Kyösti Karhila, CO of 3/HLeLv 24, Lappeenranta, July 1944

The vastly experienced 'Kössi' Karhila was made CO of 3/HLeLv 24 in place of the wounded Hans Wind on 30 June 1944. He had served with LeLv 32 (where he had scored 13 victories with Curtiss Hawk 75As) for two years, prior to being posted to LeLv 34 on 20 April 1943 – Karhila claimed a further seven kills with the latter unit. Assigned Bf 109G-6/R6 MT-461 upon his arrival at 3/HLeLv 24, he was one of the few pilots who chose to keep this version of the *Gustav* in its 'gun boat' configuration. Karhila claimed eight victories with this fighter, the last of which fell on 18 July (the final day of the the Soviet offensive) to take his tally to 32 victories from 304 sorties.

31

Bf 109G-6 (Wk-Nr 163627) MT-437 'Yellow 9' of SSgt Leo Ahokas, 3/HLeLv 24, Lappeenranta, June 1944

Ahokas was serving with LLv 32 at the start of the Continuation War, and he transferred to 3/LLv 24 on 11 August 1941. He scored seven kills with the Brewster (four of them with BW-351), before transitioning to the Bf 109G in the spring of 1944. On 4 May 1944, during the course of his conversion, he crashed MT-242 into MT-236, writing off both fighters, and killing the pilot of the latter machine. On

19 June he was issued with MT-437, but on 28 June Sgt Kosti Keskinummi wrote it off in a forced landing after suffering serious battle damage. The last of Ahokas's fighters was MT-480, which was issued to him on 7 July. He used it to down an La-5 three days later, which took his final wartime tally to 12 destroyed in 189 sorties.

32

Bf 109G-6 (Wk-Nr 167310) MT-504 'Yellow 1' of 1/HLeLv 24, Lappeenranta, September 1944

This machine was flown from Anklam to Finland on 25 August 1944, being painted with full Axis eastern front tactical markings upon its arrival. It was delivered to 1/HLeLv 24 on the final day of the Continuation War (4 September 1944), and its yellow markings were over-painted shortly thereafter as per the terms of the ceasefire.

33

Bf 109G-6/R6 (Wk-Nr 165347) MT-465 'Yellow 7' of 1Lt Atte Nyman, 2/HLeLv 24, Lappeenranta, July 1944

Atte Nyman became a member of the 2nd Flight on 2 May 1943, and he failed to claim any victories with the Brewster. He eventually 'made ace' with the Bf 109G during the summer offensive of 1944, and this particular fighter was the last *Gustav* to be assigned to him (on 28 June 1944 – he scored one kill with it). Originally configured as a 'gun boat' (one of 14 Bf 109G-6/R6s delivered), its wing cannon were soon removed by the Finns. The tactical number behind the fighter's cockpit denotes that it belonged to the 2nd Flight. Atte Nyman flew 150 sorties and scored exactly five kills.

34

Bf 109G-6/R6 (Wk-Nr 165249) MT-477 'Yellow 7' of 1Lt Mikko Pasila, 1/HLeLv 24, Lappeenranta, July 1944

MT-477 was yet another *Kanonenboote* that had its wing cannon removed. Mikko Pasila flew this fighter during the final two months of hostilities, although none of his ten kills (split equally between Brewsters and Messerschmitts), were scored with this particular machine. He completed exactly 200 sorties.

35

Bf 109G-6 (Wk-Nr 165001) MT-460 'Yellow 8' of SSgt Emil Vesa, 3/HLeLv 24, Lappeenranta, July 1944

Emil Vesa served with 3/LeLv 24 from 3 December 1941 through to war's end. Claiming 9.5 kills with Brewsters (mostly with BW-351 in 1942 and BW-357 the following year), he then transitioned onto the Bf 109G in April 1944 and enjoyed even greater success. During the Soviet summer offensive Vesa flew MT-438 until he was forced to belly-land it after suffering combat damage on 28 June 1944. He was then issued with MT-460, and went on to claim eight kills with the G-6 between 30 June and 19 July. Vesa flew 198 sorties and claimed 29.5 victories. This aircraft's assignment to the 3rd Flight was denoted by its tactical number on the fin.

36

Bf 109G-6 (Wk-Nr 164932) MT-431 of SSgt Pekka Simola, 2/HLeLv 24, Lappeenranta, August 1944

This G-6 arrived in Finland on 19 June 1944 and was immediately sent to 3/HLeLv 34, where it was damaged in a rough landing just three days later. Following a speedy repair at the State Aircraft Factory, it was handed over to 2/HLeLv 24 on 23 August and assigned to Pekka Simola. An equally new arrival on the unit, he had transferred in from MS.406-equipped LeLv 14 just too late to see any combat. The fighter's distinctive patches of black paint were a standard part of a typical Finnish Air Force camouflage scheme of the time.

37

Gloster Gamecock II (c/n 3) GA-46 of LLv 24, Utti, September 1938

GA-46 was built under licence at the State Aircraft Factory and delivered to MLE (Land-based *Escadre*) on 5 December 1929. On 15 July 1933, when Utti became LAS 1 (Air Station 1), the MLE was split into LLv 10 and LLv 24 – the Gamecocks went to the latter unit. Receiving one major overhaul during its lengthy spell with the unit, GA-46 served with LLv 24 until 26 September 1938, when it was passed on to the Air Fighting School.

38

de Havilland 60X Moth (c/n 8) MO-103 of LeLv 24, Hirvas, July 1942

Built in 1929 for civil operator Veljekset Karhumäki Oy, this Moth was impressed into the Finnish Air Force just prior to the outbreak of the Winter War. On 19 December 1940 it was delivered to LLv 24, and the biplane served as the unit's communications aeroplane until 10 July 1942, when its engine failed in flight. The Moth's pilot, Sgt Emil Vesa (a future 29.5-kill ace), successfully stalled the aircraft into a dense forest, and although both of its occupants escaped injury, MO-103 was written off.

39

VL Viima II (c/n 13) VI-15 of LeLv 24, Suulajärvi, October 1943

The Viima primary trainer was designed by the State Aircraft Factory, and this particular example was built in 1939. On 3 June 1943 it was handed over to LeLv 24 from LeLv 34, although its stay was destined not to be a long one, for on 26 October 1943 VI-15 was transferred as a 'hack' to bomber squadron LeLv 46. VI-15 is finished in a typical trainer scheme of the period. Similarly-painted VI-12 served for a much longer period with LeLv 24.

40

VL Pyry I (c/n 32) PY-33 of LLv 24, Vesivehmaa, June 1941

The Pyry advanced trainer was yet another VL (*Valtion Lentokonetehdas* – State Aircraft Factory) design. Delivered new to LLv 24 on 29 April 1941 for evaluation, the aircraft was handed over to the Air Fighting School on 5 July 1941 following the end of the trial. All advanced trainers were painted identically to frontline types.

Back cover

Brewster Model 239 BW-364 'Orange 4' of WO Ilmari Juutilainen, 3/LeLv 24, Suulajärvi, December 1942

BW-364 was assigned to 3/LLv 24's 'Illu' Juutilainen from June 1941 to February 1943, during which time he scored an amazing 28 victories with it. The fin markings show his score of two victories in D.XXIs and 34 in Brewsters up to this point. On 26 April 1942 Juutilainen became the first LLv 24 pilot to receive the Mannerheim Cross. Finland's ranking ace, he finished the war with 94 kills in total.

BIBLIOGRAPHY

PUBLISHED SOURCES (in English)

Guest, Carl-Fredrik, Kalevi Keskinen and Kari Stenman. *Red Stars.* Apali, Finland, 1995

Juutilainen, Ilmari. *Double Fighter Knight.* Apali, Finland, 1996

Keskinen, Kalevi and Kari Stenman. *Finnish Air Force 1939-1945.* Squadron Signal Publications, USA, 1998

Keskinen, Kalevi and Kari Stenman. *Aircraft of the Aces 23 - Finnish Aces of World War 2.* Osprey Publishing, England, 1998

Luukkanen, Eino. *Fighter over Finland.* Macdonald, England, 1963

Stenman, Kari. *'Finnish air force', Wings of Fame Number 12.* Aerospace Publishing, England, 1998

Stenman, Kari. *'38 to 1 - The Brewster 239 in Finnish service', Air Enthusiast 46.* Key Publishing, England, 1992

Stenman, Kari. *'Finland's Frontline - The Bf 109 in Finnish service', Air Enthusiast 50.* Key Publishing, England, 1993

Stenman, Kari. *'First and Last - Finnish Fokker D.XXIs', Air Enthusiast 88.* Key Publishing, England, 2000

PUBLISHED SOURCES (partly in English)

Keskinen, Kalevi and Kari Stenman. *Fokker D.XXI.* Hobby Kustannus, Finland, 2000

Keskinen, Kalevi and Kari Stenman. *Brewster Model 239.* Apali, Finland, 1995

Keskinen, Kalevi, Kari Stenman and Klaus Niska. *Messerschmitt Bf 109G.* AR-Kustannus, Finland, 1991

Keskinen, Kalevi, Kari Stenman and Klaus Niska. *Finnish Fighter Aces.* Apali, Finland, 1994

UNPUBLISHED SOURCES

Finnish Military Archives, Helsinki, Finland;
LeLv 24 war diary
1, 2, 3 and 4/LeLv 24 war diaries
LeLv 24 mission logbook
Combat reports
Loss and accident reports
LeR 2 operational report
LeR 3 operational report
Individual aircraft logbooks

Central Archive of the Ministry of Defence, Podolsk, Russia;
Operational reports
Orders of Battle
Loss listings

Central Naval Archives, Gatchina, Russia;
Baltic Fleet air force war diary
Operational reports
Orders of Battle
Loss listings

INDEX

References to illustrations are shown in **bold**. Plates are shown with page and caption locators in brackets.